The Lactate Revolution

The Science of Quantifying, Predicting, and Improving Human Performance

Shannon Grady

M.S. Exercise Physiologist

Testing

800m reps
10 seconds rest

Contents

Acknowledgements

Many thanks to all of the amazing coaches and outstanding athletes around the world that I have had the pleasure of meeting and working with throughout my career. Also, to my many friends, especially my University of Florida teammates, who were often the guinea pigs of "Shady's mad science" experiments. Love you Gator girls. There are too many to name personally but every one of you have played a part in making all of this possible. It would not have been possible to explore my two passions, science and sport, with so many athletes if it were not for the trust and close collaboration of numerous coaches from many sports of all levels. Special gratitude and acknowledgement to my coach at the University of Florida, JJ Clark, who also shared the same passion for wanting to enhance athletes' performances and lives in a positive way. JJ's interest in exploration of physiological data in a college setting allowed for greater understanding of the expansive possibilities of real-world applications of data into training.

To all of the Lactate Revolutionaries who came before me that sparked my love affair with lactate and its vast applications in sport. I am forever grateful for their pioneering efforts and expansive research in this space. I am also grateful for my opportunity to work with Michael Shannon at the USOC who introduced me to athlete testing and I thank him for his excellent mentorship.

I also would like to thank Ernest Maglischo on his efforts and support on reviewing my manuscript as well as encouragement to publish it. I am honored that such an experienced physiologist, esteemed coach and author gave such positive praise on my work and gave me the confidence to share my years of work with the world of sports.

Thank you to Jerry Cosgrove his continued support and exposure in the sports performance and lactate realm.

Finally, a special thank you to my son, Grady, for his many years of support while traveling the world watching his mom take endless samples of blood. Also, to my husband, Jason, who's love, support, and encouragement especially during my final push to publish this book.

My work is a culmination of years with many puzzling questions but is a product of the dedication of all the coaches and athletes involved. I hope all enjoy reading this book as a sports scientist, coach, physical therapist, doctor, athlete or anyone interested in the application of science to human performance.

I dedicate this book to my mom, Diane DeMaio, for her unconditional support and encouragement of my hopes and dreams, no matter how crazy they seemed.

Preface

When I first started out as a Sports Physiologist, I worked at the United States Olympic Training Center in Chula Vista, CA. We had many groups that benefited from our sports science resources, including resident athletes and coaches, visiting athletes and coaches, and Elite or Jr. Elite development groups. As a Sports Physiologist, my job was to provide appropriate testing and feedback on athlete's for coaches to implement into their training plans.

I quickly learned that a disconnect existed between the physiological testing and the optimal application of those results to training. After testing and evaluation, most coaches would resume their same training methods, returning a few months later, to see if their athletes had made progress in any of the testing areas. Most often, they had not.

The disconnect between testing and application was quite alarming to me. Our system to provide top athletes and coaches with the best information and methodology for optimized training responses had many weak links. Availing a dual role as a physiologist and elite athlete, I understood the implications of this disconnect on science and sport. Resolving this disconnect was the key to unlocking human potential. I knew a change was necessary. We needed a more symbiotic relationship between the scientists and coaches that would ultimately impact the success of countless athletes. After realizing there was a need for change, my quest for answers began. I began the challenge to contribute something significant to my two passions: science and sport.

I believe the primary role of a performance physiologist is to enhance the performance outputs of athletes by creating the appropriate training stimulus for each athlete to have peak performances when it counts. While there are various methods and approaches used by sports scientists and coaches to achieve peak performance objectives, my approach has always been to use a physiology first model over external or subjective measures.

All coaches and sports scientists use the knowledge gained in research, theory, application, and experience when developing their training plans to improve each athlete. After over 20 years of lactate testing, analysis, data tracking, and using a physiology first approach, I have developed a methodology that is reliable and measurable. The use of regular testing, retesting, and appropriate training application has enabled the performance results of my clients to become accelerated, repeatable, and predictable, as well as eliminate guesswork when developing training plans.

I call my testing methods and the application of its results, System Based Training (SBT). SBT evolved as a result of a culmination of years of field-studies with 100,000+ heterogenous athlete samples across all sports and levels. SBT is based on scientific evidence and not a training philosophy. The SBT testing and application uses a combination of scientific principles of lactate metabolism, lactate dynamics, applied biochemistry, applied physiology, applied exercise physiology, and nutrition. Since the early 2000s, SBT has produced numerous Olympic champions, World champions, NCAA champions, and Olympians in over 20 individual events and team sports.

The SBT testing and application provides coaches with a sound platform bound in objective data on which to develop individualized training programs that will achieve intended physiological objectives with high certainty. SBT delivers an advantageous approach to optimizing training by integrating both applied physiology and applied biochemistry to provide precise individualization of each athlete's training program. The SBT methodology has been demonstrated to provide optimized and accelerated physiological and performance responses.

When they think it can't be done, just smile and show them how :)

Chapter 1.   Disrupting What Is Comfortable

Coaching Is Art. Training Is Science

In the coaching world, longevity and experience are the commonly favored preference of expertise. This aspect of coaching is the "ART" of coaching. The "ART" of coaching deals with game strategies, specific sport skills, racing tactics, mental coping strategies, daily life strategies, and general lifestyle choices for athletic success. Coaching wisdom comes with time and experience in their respective sports and is something that is irreplaceable and serves to be an invaluable resource. The "Science" of coaching concerns the physiological responses and adaptation to stimulus, also known as training. The "Science" of coaching is simply the ability to properly apply known scientific principles in areas of human metabolism, physiology, exercise physiology, biochemistry, and nutrition.

There are numerous academic and research studies published on cellular metabolism involving physiological and biochemical processes related to performance markers such as blood lactate. Where the science of coaching is concerned there are numerous publications coaches can rely on to remain current on the best ways to train the physiological and biochemical processes involved in athletic performance. In his regard, my special interest continues to be the use of blood lactate as a biomarker

11

of training effectiveness. Still, very little information is disseminated, applied, or used in the actual daily training of many athletes. Most of the current sports science research is steered towards "proving" that invasive physiological testing is "unnecessary" to train athletes or that there are no reliable applications for blood data in training. Biomarker based physiological testing is not "necessary," but as Ben Franklin states, "in this world, nothing can be said to be certain, except death and taxes." A biomarker is a measurable substance in the blood of an organism, which indicates a physiological response. Physiological testing and athlete data is undeniably beneficial, but what is even more critical to enhancing performance outcomes through collecting data is applying that data correctly.

When training athletes, many coaches can be slow to change their methods. Change can be scary. Some coaches may resist because it may seem like more work to change, or they have had some success with their current methods. Many coaches are gaining traction and opening their minds to the world of sports science in efforts to enhance athletic performances. The emphasis on the value of invasive testing versus non-invasive testing is still lagging but growing ever so slowly through educating coaches, implementing regular cycles of testing, analyzing data, and dynamic periodization in the training of athletes based on biomarkers. Most coaches currently use various field tests in each sport, which are non-invasive methods to measure physiological functioning. Some examples of these field tests include; YIRT (Yo-Yo Intermittent Recovery Test) L1-2, RAST (Running Anaerobic Sprint Test), one or 2-mile time trials, estimates of "lactate threshold" by using 20 or 30-minute time trial or a Functional Threshold Power (FTP) test. While athletes and coaches quickly perform non-invasive tests, they are only an estimate of metabolic power or capacity of the aerobic and anaerobic systems. They are an expression of an assumed metabolic system used to perform that particular field test. Non-invasive field tests are unreliable measures of exercise intensity for most of the athlete population and have low correlations with the energy system availability to perform work for each athlete.

Why No tests

Non-invasive field tests (NIFT) are unable to measure or quantify what, why, or how the athlete is performing that test. NIFT is unable to indicate what type of training an athlete needs to improve their fitness and performances. NIFT are unable to give a coach insight as to WHY a player may pass or fail any of the non-invasive field tests. A NIFT gives external feedback with no physiological measures of what's going on inside each athlete; hence, a coach is still guessing why an athlete made it to various levels on the YIRT or improved their FTP, or conversely, why they were unable to do so. Invasive, biomarker-based field testing will give insight versus hindsight data on an individual's current physiology so that training periodization is based on actual metabolic measures rather than assumptions. The use of biomarker-based assessments is a physiology first approach. It provides unbiased insight data as to how to improve the physiology of each athlete and prescribe training that will yield a positive physiological response for each athlete.

Let me cite an example using two cyclists with the same FTP wattage targets who have to train differently to achieve optimal training adaptations and performances because other areas of energy contribution are unavailable to adapt to training or developed at different levels. Non-invasive tests are unable to measure or quantify the differences in these two athletes by using only one data point or recent performance that may be similar. However, the right type of biomarker-based physiological evaluation can.

Athlete 1:

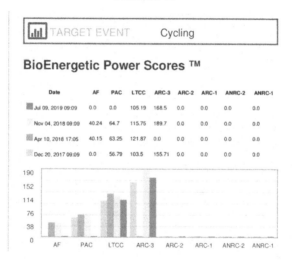

TARGET EVENT	Cycling

BioEnergetic Power Scores ™

Date	AF	PAC	LTCC	ARC-3	ARC-2	ARC-1	ANRC-2	ANRC-1
Jul 09, 2019 09:09	0.0	0.0	105.19	168.5	0.0	0.0	0.0	0.0
Nov 04, 2018 09:09	40.24	64.7	115.75	189.7	0.0	0.0	0.0	0.0
Apr 10, 2018 17:05	40.15	63.25	121.87	0.0	0.0	0.0	0.0	0.0
Dec 20, 2017 09:09	0.0	56.79	103.5	155.71	0.0	0.0	0.0	0.0

Athlete 2:

BioEnergetic Power Scores ™

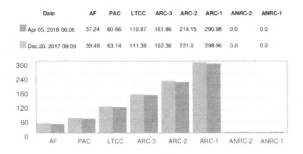

Date	AF	PAC	LTCC	ARC-3	ARC-2	ARC-1	ANRC-2	ANRC-1
Apr 05. 2018 06:06	37.24	60.66	110.87	161.86	214.15	290.98	0.0	0.0
Dec 20. 2017 09:09	39.49	63.14	111.39	162.36	221.0	298.96	0.0	0.0

Certainly, biomarker-based physiological testing and evaluations are unable to guarantee every athlete will become an Olympian or World champion. Still, they are a huge step forward in the progress and advancement in human performance optimization.

The "Art" of coaching requires hard work, dedication, and passion for being successful, but the science of training athletes involves knowledge of human physiological responses to various physical stimuli. Figuring out what "works" for each individual without physiological data can be challenging. It can take months or even years of trial and error training to make performance gains in an athlete. Artful decision making remains a necessary component of good coaching. After all, athletes, even of the same age and sex will respond to training differently, therefore, even the most valid scientific tests will be 100% accurate. That said, the data gathered from regular use of valid scientific testing that is based on the most current physiological will greatly increase both the number of athletes who improve performances in a particular

15

season and the amounts by which they improve. To be sure, regular testing that is based on the most accurate and up to date research can significantly benefit any coach or athlete looking to optimize performance and eliminate guesswork when individualizing training. What follows is a response from one highly successful track coach who put System Based Training to the test.

"I approached Shannon Grady because I had an extraordinary athlete that I wanted to make sure he was receiving the appropriate training at any given time of year. I wanted the guesswork of what to do when and how fast or slow taken out of the equation. He was a once in a lifetime athlete, and I only had a limited amount of time to get it right for him. I had a professional relationship with Marcus O'Sullivan, the head coach at Villanova University, who was a client of Shannon's who had much success. So, I decided to reach out to Shannon to see if she would be willing to help me. She couldn't have been more receptive and accommodating. After several months of working closely with Shannon, the result was my athlete improving from 1:49 for 800 meters to 1:46.70 and becoming runner up at the 2015 NCAA Indoor Track and Field Championships. Shannon was readily available to help me with the overall concept of what we were doing and eager to answer any questions that arose. The experience was so successful and so enjoyable that the following year, we decided to implement her System Based Training methods with the entire program and have seen similar improvements across the board, both male and female. I would recommend Shannon Grady to not only inexperienced coaches

just getting started in their coaching career, but also to experienced coaches looking to try to take their programs to a new level."

- Chris Tarello - Monmouth University Associate Head Cross Country - Middle Distance Track & Field

Chapter 2. The Lactate Revolution

"Lactate is the key to what is happening with metabolism, that IS the revolution."

- George Brooks

Fallacies of Traditional Lactate Threshold Testing and Training

Many applied sports scientists and coaches use various applications of lactate testing in daily training. However, many academic researchers have concluded that lactate testing is disadvantageous for training athletes because the results of academic research is often disassociated with the "real world" of athlete training. For example, the design of academic studies is often short-term observations of a small, homogeneous athlete group with the hopes of making correlations and significant findings compared to pre-test data or control groups. Most often, the athletes in the control groups are labeled "well-trained," but in actuality are in a state of physiological dysfunction so the stimulus applied to the group of athletes will not create significant physiological changes. In addition, in most academic study designs, dynamic physiology is not taken into account. Unfortunately, this stance in academia has permeated the training literature as there are few books on training that cover the topic. In the confines of many academic studies, the limitations make it clear as to why academia has this stance. It is very challenging, and I would even go as far as saying impossible, to understand the implications individual lactate dynamics have on human performance, especially in non-endurance events, in a controlled, laboratory setting. It is a far stretch of the imagination to assume that concepts from a limited protocol in a controlled laboratory setting will apply to the dynamic conditions of training and competition in numerous athletes. Most often, academic laboratory studies are observations of less than ten homogeneous subjects over less than four weeks.

Much of academic research studies on lactate are on endurance sports athletes who will give a limited perspective on the dynamic nature of lactate in human physiology. There are several reasons why the complexity of using lactate measurements as a reliable marker for performance and training has failed in many academic studies. Since lactate kinetics vary for each individual, a slight change in the testing protocol or data collection methods will complicate outcomes of such data and make data from different protocols impossible to compare. The inferences from lactate data are limited to specific protocols which should be repeatable from test to test to have reliable data and applications. Different protocols are unable to generalize to others' results, but amongst sports scientists, academics, and coaches, no one agreed-upon protocol is the gold standard. The lower net lactate values that indicate a high aerobic energy contribution is most available and unchanged when tested, giving a false perspective of the true dynamic nature of lactate and human physiology. Most study designs are looking at and a limited view of the human energy spectrum as

19

well as a similar athlete group as their sample size. Lactate testing is unable to answer everything about one's physiology but using lactate data as an objective approach to planning all aspects of training can be highly beneficial when done correctly.

Some athletes will undoubtedly improve without using lactate testing in their training program, nevertheless, the data that results from the tests used in System Based training has proven to be absolutely, and undeniably advantageous to enhancing the performances of athletes of both sexes, different ages and widely varying abilities. Even more important than collecting physiological data is applying that data appropriately. Application of lactate data to improve performance in real-world athlete scenarios can take years to begin to understand, as human lactate data is exceptionally dynamic. The analysis of lactate data should be results from repeatable protocols over several years to be able to interpret and implement in the daily training of athletes.

Training based on measurement of the "lactate threshold" has become very popular during the last two decades but it has its limitations. The "lactate threshold" is often defined as the maximum effort that can be sustained for an "extended" period without lactate steadily accumulating. This definition and term are very ambiguous with really no agreeable defined parameters across scientific literature.

I abstain from using the term, lactate threshold, when discussing lactate, lactate dynamics, or lactate application in training for several reasons;

1. There is no one training speed that can accurately be called the single lactate threshold. Instead, there are several such thresholds associated with the many different aspects of aerobic and anaerobic metabolism that comprise the energy metabolic continuum.

2. Each net lactate threshold is trainable and needs to be reevaluated via blood lactate testing regularly.

3. There is not a sudden shift from aerobic to anaerobic metabolism at a net blood lactate reading of 4 mmol (the most frequent determination of "lactate threshold").

4. The emphasis of training design to enhance the "lactate threshold" is often responsible for bioenergetic deficit and underperformance.

Traditionally, measures, methods for determining a single "lactate threshold" (commonly called anaerobic threshold or OBLA) or VO2 max values have been the gold standard as predictors of performance and also to develop training zones and recommendations for increasing aerobic and anaerobic capacities.

Unfortunately, the significance of emphasizing "lactate threshold" or VO2 max values as the determinants for daily training will lead to an imprudent and hasty decisions based on physiological data which does not include the many and varied aspects of aerobic and anaerobic energy metabolic continuum that play important roles in athletic performance. The "lactate threshold" and VO2 max values are metrics that fail to develop individual physiology when used in daily, monthly, and overall training plan development.

Most scientists, coaches, and athletes quite often think of "lactate threshold" as the point in which energy shifts from primarily aerobic to anaerobic energy sources. This concept of energy shifting at this point of net lactate is not truly an aerobic to anaerobic change in most humans. The human physiological capacity and energy shifts to anaerobic sources occur at much higher net lactate output levels for most athletes. Conventional estimates of "anaerobic threshold" are typically about 50% lower than actual anaerobic energy shifts. Most of the standard physiological assessment protocols involving the use of lactate data neglect to evaluate the entire range of the energy metabolic system. Most protocols stick with the textbook "4 mmol" value as the "anaerobic threshold." The fact that human physiology is dynamic is the fundamental problem with basing training percentages, wattages, or paces on a single net lactate value.

In the real world of actual human physiological profiling, a net lactate value of 4 mmol can be anywhere from 16 to 100 percent of one's maximum net lactate.

Physiological profile analysis of 250 Division I middle distance runners who were underperforming, most often required a weakening or slowing down of their textbook defined "lactate threshold" or the speed at four mmol values to improve performances over 800 to 3000 meters. In sports science literature, there is no consistency as to defined parameters "lactate threshold" training actually is or how it is determined. Some of the definitions can be as vague as "anaerobic threshold" or "increases in lactate." Those who seek to train only at the "lactate threshold" are missing the big picture that encompasses the entire spectrum of bioenergetic systems involved in training. They fail to realize that what most think of as "lactate threshold" workouts will be different for each athlete.

Many field tests to determine "lactate threshold" consist of having athletes perform protocols such as "Functional Threshold Power," (or "FTP"), or a 20 to 30-minute time trial. For most trained athletes, typical "lactate" threshold or FTP field testing protocols will lead to under stimulation of their true bioenergetic potential. On the contrary, for many untrained, glycogen depleted, or bio energetically imbalanced athletes, the typical "lactate" threshold field testing protocols will lead to overworking their actual physiological limits. These short-sighted methods will lead to further bioenergetic imbalance and underperformance. The notion that all people have an "anaerobic" or "lactate" threshold of 20 to 30 minutes is not an accurate representation of physiological capacities for these net lactate outputs. What most consider "lactate threshold" can range from a capacity of 1 minute to 75 minutes depending on the individual's current physiological status and physiological profile.

Unfortunately, in endurance training, there is an overemphasis on "lactate threshold" training and lack of emphasis on maximizing an individuals' physiology. This approach will undoubtedly limit human power and capacity limits. Without the actual knowledge of individual physiological system development and capacities, lactate threshold training designated as velocity that produces a blood lactate of 4 mm/l or even with calculations of an individual lactate threshold there are several unexpected consequences that may occur. Unexpected consequences such as, performance plateaus, chronic glycogen depletion, and what is commonly referred to as a state of overtraining. Training "lactate threshold" too frequently has proven to be, in my experience, to be very detrimental to an athlete's physiological profile and performance outputs, especially in events shorter than 90 minutes. Various academic studies have also concluded that training too frequently at the "lactate threshold" can overstress the sympathetic nervous system, slow recovery, and decrease glycogen utilization.

LTCC or "lactate threshold" training is an essential and critical area in the overall improvement of almost every athlete. Still, the volume, intensity, and frequency of this type of training needs to be given more significant consideration when planning. A key concept in the application of training using lactate data is that only focusing on one area of physiology or workout type, such as training too frequently at the lactate threshold, can prove detrimental to physiological development and performance improvements. On the other hand, training that includes training for the entire spectrum of energy metabolism

24

from highly aerobic to highly anaerobic are valuable and useful as long as that stimulus is appropriate for that athlete at that time. Overemphasis of any System or workout type will undoubtedly lead to underperformance and bioenergetic deficits.

Many mid-distance athletes (800-1500 meters) and team sport athletes have a higher percentage of fast-twitch or Type II muscle fibers. Fast-twitch (glycolytic) muscle fibers are associated with anaerobic actions such as high power and explosive energy, but they fatigue faster than slow-twitch or Type I (oxidative) muscle fibers. Slow-twitch fibers are associated with aerobic actions such as long endurance events. Fast-twitch fibers will use glycogen as their primary energy source versus slow-twitch fibers, which use primarily free fatty acids and blood glucose. Successful mid-distance athletes will have a genetic predisposition for stronger anaerobic systems. Consequently, training too frequently at or near the "lactate threshold" is a significant contributor to decreases in performances in middle distance events, especially during peak season. These athletes, in particular, get little or no improvements in performance with an emphasis on "lactate threshold" training or even increases in "lactate threshold" pace. So, as a coach of events shorter than 60 minutes, is it best to focus on all bioenergetic system power and capacity improvements rather than focusing on improving the "lactate threshold." The following statement by Vaughn Craddock, a highly successful elite track coach from New Zealand offers support for the statements I have just made.

25

"In both my roles as a physiotherapist (physical therapist) and a running coach, I am constantly searching for the optimal balance between "the science and art" of coaching to maximize the training/treatment effect.

When I first started as a coach, I had a solid scientific background from my job as a physio, and I had a wealth of real-world experience through having been an elite athlete myself. But I had no clearly defined way to merge those two elements successfully.

After two very successful years of coaching, when one of my athletes started to regress with no apparent reason, I realized I still lacked the "missing link" to help give clarity to the question of how I could control the training variables to ensure more reliable and repeatable gains in performance.

This led me to search out leaders in their respective fields and pick their brains for insights. Through this process, I was introduced to Shannon, which has subsequently proved to be a turning point for my coaching.

Because of Shannon's mentoring, I have been able to integrate the SBT methodology into our training squad, resulting in far greater reliability in performance outputs.

The real-time biofeedback that the PPT testing provides, in conjunction with the practical scientific insights of SBT methodology, has allowed me to better understand the complex interactions of both subjective and objective variables in my athletes' training. These insights lead to greater confidence in decision making, and the ability to proactively rather than reactively adapt an athlete's programming. This proactive approach, rather than the ambulance at the bottom of the cliff approach we are sometimes forced to take if we don't adequately understand the many varied, unique and individual needs of each athlete, has led not only to rapid improvement in performance, but also a significant reduction in unplanned time off due to injury or illness.

For anyone searching for a tool to help them better understand their athlete's unique needs and to optimize their performance better, I can confidently say by accessing PPT testing and integrating SBT methodology into your training/coaching systems you will take a dramatic step forward."

-Vaughan Craddock- Physiotherapist and Elite/Pro Track Coach New Zealand (High Performance)

After implementing SBT, every athlete in Vaughan's elite running squad had achieved at least one lifetime PB, and in many cases, they had produced multiple PBs. One of those athletes broke a National record that had previously stood for a significant time. From the pool of 35 track and field athletes who have undertaken PPT testing in 2018 in New Zealand, there have been 28 Personal bests, 1 National Record, and 1 Resident National Record achieved.

Lactate Revolutionaries

While there are over 5000+ published articles on the various uses of lactate testing in sport, here is a brief overview of some pertinent scientists and their research that have inspired my career and fascination with the application of lactate testing to performance. My work and interest in the use of lactate in the evaluation and prescribing of training for athletes began in 1997. After spending about four years of laboratory research on cellular metabolism, in particular, exercise biochemistry, in animals and humans, I started reading some of the work of several well-known scientists. Three of my favorite Lactate Revolutionaries in the last 50 years are Alois Mader, Dr. Bruce Gladden, and Dr. George Brooks.

Mader was one of the earliest scientists who studied lactate and performance. Mader's research sparked the revolution of the use of lactate in human performance testing and analysis. Mader evolved many training theories found in his research and writing, along with Jan Olbrecht, who was one of Alois' colleagues. Mader and Olbrecht suggest regular blood lactate testing in swimmers to monitor and prescribe training intensity, as well as quantify changes in the aerobic and anaerobic energy systems as a method to evaluate training effects. Olbrecht's methods generally involve applying the use of lactate testing to individualize and adapt the stimuli to each athlete in hopes of maintaining optimal aerobic and anaerobic energy balance for optimal performances.

Mader and Olbrechts' findings and methods have been applied to many world-class swimmers and triathletes. These two Lactate Revolutionaries are best known for research on training theories. Still, their application of these findings is rarely discussed in training publications. Surprisingly, some experts have even attempted to discredit their methods due to the lack of universal use for all athletes. However, as Jan Olbrecht points out,

"The keys to success do not lie in training hard but in training purposively and carefully. Failing to plan and follow a training plan and competition program will surely undermine your athlete's potential and inevitably result in frustration or, what is even worse, in overtraining or injuries"

- Jan Olbrecht

Mader and Olbrechts' application of lactate testing to training has an emphasis on the importance of glycolysis or breakdown of carbohydrate molecules, the lactate production process, and how it can be used trained with athletes in both sprint and endurance events. Until the mid-2000s, Jan Olbrecht was implementing the use of various lactate testing protocols for daily training prescriptions with more coaches and athletes than anyone else in sports. Jan has produced many Olympic and World champions in swimming and Ironman races. Jan has primarily limited his work to swimming and endurance sports. I became quite interested in Jan's application and used it in daily training, but I shifted my focus of testing and implementation to primarily anaerobic events and team sports while developing my approach to System based training for all sports.

Alois Mader, Dr. Bruce Gladden, and Dr. George Brooks have produced many in-depth articles on lactate research from lactate metabolism and theory to the application of lactate testing to training. The area that was of particular interest to me was its application to the more anaerobic events and team sports. In this respect, George Brooks' concept of the "lactate shuttle" proved to be particularly illuminating. George Brooks is best known for his work, "The Science and Translation of Lactate Shuttle Theory," which I will review in detail as it relates to performance in Chapter 5.

Any review of lactate research and training would not be complete without mentioning the work of Dr. Bruce Gladden. Gladden's work focused on the in-depth scientific details of the cellular metabolic processes that occur during exercise. A few other publications that may help one understand these applications to exercise and physical activity are those of Perry, Heigenhauser, Spriet, and Howlett.

Several Europeans have also contributed substantially to the knowledge of the cellular processes involving muscle fibers and enzymes in aerobic and anaerobic metabolism: In his book, Cellular Metabolism and Endurance (1992), noted researcher, J. Henriksson, from the Karolinska Institute in Stockholm, has published numerous in-depth articles as a basis for understanding the science of various aspects involving performance and lactate dynamics. Henriksson discusses his and others' research on muscle fibers and the enzymes associated with energy production. He states that "If the body has higher amounts of aerobic enzymes, then it will be able to produce aerobic energy more efficiently. It is also prevalent to associate changes in the aerobic system from training and detraining with changes in enzymes." There is little about lactate per se in this article, but it provides a layman's discussion of enzymes and how they respond to training and how net lactate levels are determined by the various enzymes in the anaerobic and aerobic energy systems of the muscles. There is also some discussion of how multiple types of training will affect enzymes. Still it should be said that, despite the work of Henriksson and a few others, the application of various training stimuli and their effects on individual lactate dynamics in athletes in published research is almost non-existent.

One theory Henriksson discusses, which I have tested at great lengths, is how the anaerobic system can be trained or de-trained. Training is intended to enhance a particular area of physiology, while de-training is, in essence, "turning off" or making unavailable a specific area of physiology. I applied his theory across a heterogeneous set of athletes in various physiological states. One conclusion I came to is that training the anaerobic system is very dynamic, complex, and can "turn off" rather quickly with inappropriate training stimulus. Fortunately, the anaerobic system can be managed easily with insight from lactate data on an athlete's actual and current physiological state of bioenergetic availability. Having insight on an athletes' current anaerobic system is a key to performance outcomes in events shorter than sixty minutes in duration and in virtually all team sports.

Another critical lactate-related topic when it comes to team sports and uniquely individual sports events shorter than ten minutes in duration, concerns the function of something called lactate transporters. Another name for lactate transporters that exists in the literature is monocarboxylate transporters (MCTs). An understanding of how lactate transporters, work is fundamental to understanding how to train appropriately for events that require contributions from both the anaerobic and aerobic energy metabolic systems. Lactate transporters are proteins that carry lactate across cell membranes and play an essential role in the pH regulation of skeletal muscle. MCT expression occurs in muscle fibers and can occur rapidly. MCTs are also bi-directional, which means that they can both release and take-up lactate depending on the pH gradient of tissue to the surrounding environment. The fact that the concentration or pH gradients drive lactate shuttles is a fundamental convention I used in the development of my HMCT ™ System training approach. While most sports scientists and coaches have never heard of this term, I have mastered training this aspect of lactate dynamics, via regular testing and monitoring, in countless World Class, Elite athletes.

I recommend reading *Lactate-Proton Cotransport in Skeletal Muscle and Lactate Exchange,* as well as *pH Regulation in Skeletal Muscle* by Carsten Juel of the August Krogh Institute at the University of Copenhagen (Juel, C. 1997, 1998). Both

32

technical reviews present two key findings: 1) that lactate and hydrogen ions move together across the cell membrane via a series of lactate transporters, and 2) the transporters that do this are trainable. Training of lactate transporters in athletes requires workouts that upregulate MCTs and also address lactate shuttling via aerobic metabolism. HMCT™ System workouts will necessarily decrease the lactate concentration of the working muscles and shuttle it to other areas for recycling or removal. Important performance implications of HMCT™ System workouts are that they enable athletes to race or compete at a higher rate of velocity longer. Upregulation and training of MCTs is essential for peak performance in events that are shorter than ten minutes and those that require the use of repeated anaerobic and alactic efforts.

"The Science and Translation of Lactate Shuttle Theory"- An Overview Of George Brooks' publication

Lactate has primarily been associated with and referred to as a metabolic waste product involved with fatigue during exercise. Lactate was once thought to accumulate as a consequence of anaerobic conditions, or lack of oxygen, in working muscles. However, it was later discovered that lactate is also formed and utilized continuously, when oxygen is present, during aerobic work.

In **Figure 1**, "The Lactate Shuttle Concept," as depicted by Brooks, displays lactate as the link between glycolytic (anaerobic) and aerobic pathways. According to Brooks, the lactate shuttle hypothesis, which describes the linkages between lactate "producers" and "consumers," exist within and among cells, tissues, and organs with lactate as the crucial link. Brooks, demonstrated lactate production continuously occurs under fully aerobic conditions and can transcend compartment barriers and occur within and among cells, tissues, and organs; hence, the "shuttling" of lactate throughout the body.

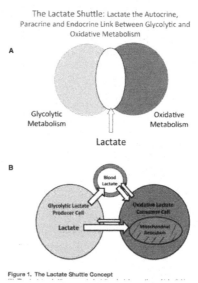

Figure 1. The Lactate Shuttle Concept

Figure 1. The Lactate Shuttle: Lactate the Autocrine, Paracrine, and Endocrine
Link Between Glycolytic and Oxidative Metabolism

Both Brooks and Gladden agree that short-term challenges to adenosine triphosphate (ATP) supply stimulate increases in

34

lactate production, leading to immediate, short, and long-term cellular adaptations to support ATP homeostasis. ATP is a molecule that carries energy within cells and is the primary energy currency of the cell. All living things use ATP to produce the energy for work.

Homeostasis is the ability of a living organism to maintain a balance of internal physical and chemical conditions. I was able to observe this critical concept in many athletes. Analysis of hundreds of physiological profile tests, stress hormones, blood chemistry panels, nutritional analysis, and performance data displayed that limitations of one's lactate production, lactate shuttling, and lactate clearance were all crucial aspects needed to support and maintain overall homeostasis and also peak performances.

Figure 2 As depicted by Brooks, has two main concepts "Cell-cell lactate shuttle" and "Intracellular lactate shuttle". It illustrates the role of lactate in the delivery of oxidative and gluconeogenic substrates as well as in cell signaling. Gluconeogenesis is the term for the metabolic process that involves creating new glucose from non-carbohydrate carbon substrate sources such as lactate. The term "Cell-cell lactate shuttle" includes the shuttling or exchange of lactate between white-glycolytic (Fast-Twitch, Type II) muscle fibers and red, oxidative (Slow-Twitch, Type I) muscle fibers within a working muscle. The term "Intracellular lactate shuttle" is for describing the shuttling or exchange of lactate between working skeletal muscle and the heart, brain, liver, and kidneys.

Figure 2. Depiction of the Lactate Shuttle as It Fulfills Three Physiological Functions Lactate is a major energy source; is the major gluconeogenic precursor; is a signaling molecule with autocrine- paracrine- and endocrine-like effects; and has been called a "lactormone." "Cell-cell" and "intracellular lactate shuttle" concepts describe the roles of lactate in delivery of oxidative and gluconeogenic substrates as well as in cell signaling. Examples of the cell-cell lactate shuttles include lactate exchanges between white-glycolytic and red-oxidative fibers within a working muscle bed and between working skeletal muscle and heart, brain, liver, and kidneys. Examples of intracellular lactate shuttles include cytosol-mitochondrial and cytosol-peroxisome exchanges Indeed, most if not all lactate shuttles are driven by a concentration or pH gradient or by redox state. G, glucose and glycogen; L, lactate. Compiled from diverse sources. [Brooks, 1984, 2002, 2009].

buffers ([A⁻ₜₒₜ], in plasma mainly amino acids and proteins], and the strong ion difference (SID), which is calculated as

Figure 2. Depiction of the Lactate Shuttle as it Fulfills Three Physiological Functions

Lactate metabolism is now understood to be essential for at least three reasons:

1. Lactate is a primary energy source

2. Lactate is the major gluconeogenic precursor

3. Lactate is a signaling molecule with effects similar to autocrine, paracrine, and endocrine molecules.

Another key finding of Brooks and others that should be applied to the training of lactate shuttling is that potentially all cell-cell and intracellular lactate shuttles are driven by a concentration (pH) gradient or by a redox state. The rapid exchange of lactate shuttling across cell membrane barriers is possible via MCT proteins. Lactate shuttling is vital in brain functions in healthy, as well as pathophysiology, or disordered physiological processes associated with various disease states.

Lactate exchanges between blood, producer (glycolytic, Fast-Twitch, Type II fibers) and consumer (oxidative, slow-twitch, Type I) cells refers to the "cell-cell lactate shuttle" within a tissue bed that moves down lactate and proton concentration gradients, in other words, from areas of high lactate concentration to regions of low lactate concentrations.

During continuous, physical exercise, it is important to note an extremely crucial concept in order to understand the physiological implications of blood lactate data, that lactate concentrations found in the blood are a net value. The term net lactate value means it is the combined effect of both the simultaneous production and elimination of lactate during and after exercise. Net blood lactate concentrations will be an intermediate value that is higher than the consumer, oxidative cells, but lower than the producer, glycolytic cells. It is imperative to understand and consider this concept when evaluating the physiological effects of a recent training stimulus has had on an athlete from post-exercise measurements of net blood lactate.

"There is strong evidence that glucose and glycogen catabolism proceed to lactate production under fully aerobic conditions."
-George Brooks

The main points from Brooks' review on the physiology of glycolysis are that:

1. Glycolysis makes lactate.

2. Lactate is a necessary fuel and signaling molecule. Therefore, it is critical to consider the physiological factors that stimulate lactate production.

3. Traditional assumptions have been that rising lactate levels indicated anaerobic conditions at the cellular level.

4. However, current research is, as stated earlier, that some lactate is also produced under fully aerobic conditions.

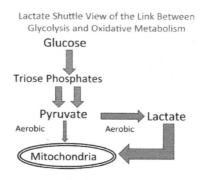

Lactate Shuttle View of the Link Between Glycolysis and Oxidative Metabolism

Figure 3. Brooks' Depiction- Glycolysis Produces Lactate, Not Pyruvate

"Characterization of lactate as a dynamic fuel, produced even when oxygen is not lacking, redefines this molecule from being an apparent nuisance physiologically to being of metabolic benefit. Scientific views have evolved over the years regarding the role of lactate in causing muscle fatigue."

-George Brooks

The main points from Brooks' review on lactate and fatigue are that:

1. The traditional conclusion has been that lactic acid causes muscle fatigue because of the central role of lactate in intermediary metabolism and exercise performance.

2. This was because previous research sought to distinguish between the separate effects of lactate anions and hydrogen ions on muscle performance.

3. It is appropriate to conclude that exercise causes both lactate anion and proton accumulation.

4. It is unclear whether glycolysis makes lactic acid or if accumulated lactate anions and protons are derived from the same sources.

5. It is also unclear if lactate and hydrogen ions, individually, separately, or in aggregate, are causes of muscle fatigue and that lactate is the single cause of fatigue.

6. Physiological states are affected by protons from more than one metabolic pathway, and protons from glycolysis do not affect blood pH on a 1:1 basis.

7. Recent reviews support the conclusion that the combined effects of acidosis, phosphate ion accumulation, and low Ca 2+ interfere with muscle cross-bridge cycling and performance.

8. Consequently, at present, it is uncertain that proton accumulation during human muscle exercise is a significant cause of muscle fatigue.

9. We know that glycolysis, the breakdown of glucose is a means to produce ATP that allows cellular work to progress.

10. In various studies it has been concluded that glycolysis is pH neutral because it produces lactate anions, as opposed to lactic acid.

"In summary, in this section, after a century of effort, what can we conclude on the interrelationships among cell work (e.g., muscle contraction), glycolysis, lactate, and proton accumulation? We can conclude that cell work stimulates glycolysis and lactate anion and proton formation, but there is persistent controversy over whether glycolysis produces lactate or lactic acid. Further, there is a similar controversy over whether either lactate anion or proton accumulation interferes with the mechanism of muscle contraction and causes fatigue. Certainly, glycolysis is necessary for muscle power generation, and certainly, also lactate provides a fuel energy source."

- George Brooks

The main points from Brooks' review on the energetics of lactate production are that:

1. Glycolysis is the entry pathway for catabolism, or breakdown, of carbohydrate products via digestion.
2. The oxidative catabolism of a carbohydrate molecule yields 4 kcal/g, or 34-36 mol ATP/mol glucose.
3. Glycolysis involves both sustained, oxidative, and short-term, glycolytic high muscle power output activities.
4. There are two modes of glycolytic function during exercise: (1) continuous glycolysis with lactate production (Ra) matched by oxidative disposal (Rd) and

(2) accelerated glycolysis in which lactate Ra far exceeds the capacity of lactate Rd via oxidation.

5. Sustained muscular power output depends on highly integrated processes involving cardiopulmonary, cardiovascular, and oxidative muscle capacities.

6. Athletes that generate higher muscle power output are due to the high capacity for muscle glycogenolysis and glycolysis.

7. Muscular power and endurance require an individual to sustain glycolysis and lactate clearance without incurring hyperlactatemia.

8. In non-endurance and high-power sports, muscular power comes from glycogenolysis and glycolysis. Glycogenolysis is the breakdown of a glycogen molecule and glycolysis is the breakdown of a glucose molecule.

Chapter 3. The Science of Lactate Dynamics

Net lactate outputs are an objective, quantifiable, and reliable biomarker that has a direct correlation to an athlete's ability to perform in any sport or event.

Understanding lactate dynamics is a crucial component to unlocking human potential. The lactate ion, C3H5O3, is the conjugate base of lactic acid, C3H6O3. Although these two compounds are interchanged when referred to amongst the masses, they are indeed two unique molecules. The chemical difference between lactate and lactic acid is that lactate is lactic acid, missing one proton (H+). To be considered an acid, a substance must be able to donate a hydrogen ion; when lactic acid donates its proton, it becomes its conjugate base or lactate. The human body produces and uses lactate, not lactic acid. Humans have a spectrum of possible net blood lactate values that ranges from 0.7 mmol to 32 mmol. Net blood lactate values are often dismissed as an unreliable data point due to the dynamic nature of this complex, yet fascinating molecule.

To date, the easiest, most convenient way to measure net blood lactate values in the athletic environment is by finger or ear sampling. Although, collecting blood lactate samples via portable lactate analyzers may seem simple, due to the dynamic nature of the lactate molecule, the sample collection must and should be done using very precise and repeatable protocols.

No spot checks

Random or spot check sampling during a workout or training session is standard practice by coaches to estimate work intensity but this method is the least scientific approach as it is unreliable and provides invalid data points when "studying" lactate. A random sampling of blood lactate neglects to give any indication of changes in fitness, effort levels, intensity or how hard an athlete is working during any particular workout. The dynamic nature of lactate kinetics variability, within and among individuals, will dramatically affect net readings at varying intensities of work and rest periods during interval training.

The four main components of lactate dynamics are lactate production, lactate elimination, lactate buffering, and lactate shuttling. Lactate production is the result of the breakdown of glycogen or glucose. Lactate elimination is the process of removing lactate from the muscles and bloodstream. Lactate buffering is the ability of the body to neutralize part of the acid or H+ ions produced from lactate accumulation. Lactate shuttling is the process that moves or shuttles lactate around the body.

In studying lactate data in athletes, it is imperative to keep in mind the underlying physiology and biochemistry of this fantastic molecule and all of the variables that influence lactate production, elimination, buffering, and shuttling. Although adaptation, an athlete's lactate dynamics, is highly trainable and predictable. A fundamental concept when observing changes in lactate data is that a variety of factors influence lactate dynamics in addition to the training stimulus. Lactate dynamics can also be affected by nutrition, illness, disease, or other metabolic limiters. Unfortunately, net blood lactate readings are incapable of indicating which aspect of lactate dynamics has been affected by training, nutrition, illness, disease, or other metabolic limiters but will give an accurate depiction of one's current bioenergetic availability. Without long-term and repeated testing throughout various training periods of a large data set of a heterogeneous sample of athletes, interpreting lactate data may be challenging. Grasping human lactate dynamics will also require an in-depth knowledge of physiology, exercise physiology, biochemistry, nutrition, and metabolism.

Lactate production occurs under aerobic and anaerobic conditions as a product of glycolysis or glycogenolysis. Mainly, lactate production occurs at all times at some level. It is important to note, changes in net lactate values, especially maximum lactate and velocity at similar net lactate values from test to test, are most often due to limited carbohydrate consumption or glycogen stores and inappropriate training prescription rather than changes in fitness levels.

The following factors mainly influence lactate elimination or clearing:
1. Aerobic enzymes or molecules of protein that act as catalysts in the steps of metabolism.

2. Number and size of mitochondria, also known as the powerhouse of the cell. Mitochondrial density plays a significant role in improving aerobic energy contribution therefore improving lactate elimination and sparing early onset anaerobic energy contribution. Sparing anaerobic energy will delay fatigue in sports that require repeated anaerobic energy contribution or events that are continuous in nature even as short as 30 seconds.

3. Capillarization or formation of new capillaries

Increasing one, two, or three of these factors can significantly enhance lactate elimination rates, which therefore can improve muscle endurance, velocity, power, and capacity. Adequate lactate elimination is a crucial component during a performance in any sport or event. Lactate elimination is unable to make lactate disappear entirely from the body but does allow it to be dispersed to areas of lower lactate concentration during work.

LO CARB
NO

Contrary to some popular dieting suggestions, enhancing aerobic enzymes and reliance on fats as fuel fails to be stimulated by eating more fats and eliminating or dramatically decreasing carbohydrate intake. To the contrary, an increased reliance on fats and aerobic metabolism can be accomplished by applying the appropriate training stimulus and will not be enhanced by eating more fat, which is most often adequately consumed in the diets of most people. At the same time limiting the daily ingestion of one's carbohydrates will almost certainly limit an athlete's ability to train, benefit from such training, and increase metabolic adaptability (discussed further in Chapter 11.

Lactate, produced in one set of muscles, can be consumed by other muscles, principally slow twitch fibers. It can also be consumed by the liver, or the heart. When a muscle consumes lactate, it will most likely be used for aerobic energy. If the liver absorbs the lactate, it will be converted back to glucose or glycogen and eventually undergo glycolysis or glycogenolysis.

Lactate is the epitome of recycling, a miraculous molecule that keeps on giving.

The two parts of lactate dynamics that are critical to performance, especially in shorter, higher energy demand events or sports, are lactate buffering and shuttling capabilities. An athlete's ability to buffer and shuttle lactate will enable athletes to perform at a higher velocity and/or higher power outputs repeatedly for longer periods.

The body can also develop a finely tuned process to buffer lactate or neutralize hydrogen ions in order to maintain homeostasis during exercise. Lactate can accumulate in various quantities depending on the time, intensity, and duration of an activity. During exercise, hemoglobin carries oxygen from the lungs to the working muscles. Working muscles have a higher oxygen demand than at rest so the muscles will become depleted of oxygen. Muscles will produce increased amounts of carbon dioxide and hydrogen ions and then they flow into the blood. At the same time, hemoglobin in the blood will pick up extra hydrogen ions and carbon dioxide to aid in buffering and maintaining blood pH. The buffering capacity of hemoglobin is easily exceeded but fortunately, excess hydrogen ions can also be removed by the kidneys and lungs which will provide a faster way to control the pH of the blood.

To reiterate, lactate shuttling, which I have discussed while reviewing Brooks' work, is possible via lactate transporters, namely monocarboxylate transporters (MCTs). The lactate shuttling concept is fundamental to an understanding of training athletes, especially in events that require a contribution from the anaerobic and aerobic systems.

One other term that is important to mention because it is often interchanged with lactate buffering, is lactate tolerance. Lactate tolerance is the ability of a person to withstand increased lactate levels in the muscles. Lactate tolerance is primarily a mental training exercise as opposed to lactate buffering which is a change in the biochemical state, as mentioned above.

Although all four main components that affect lactate dynamics can be influenced by variables other than training, it is important to understand that all four components are also trainable and, in fact, may be more amenable to the proper balance of a variety of various types of training than they are to any other single factor. The specifics of how to train each of the four factors from production, elimination, buffering, and shuttling will be discussed in Chapter 5. Following is a testimonial from a satisfied user of System Based Training.

"As far as testimonials go, I'd say our whole season was a testimonial to the benefits of lactate testing. We were able to "over-achieve" as a team. We had lost 6 of our top 8 runners from last year, and yet we ended the season with conference and county championships, a #1 ranking in South Jersey, and a 6th place finish at the NJ State All-Group Meet of Champions. None of which was predictable based on the previous year's results. I should note that two of our 800m runners have benefited tremendously from the more appropriate training resulting from the lactate testing. Rich Nelson and Will Andes were our 3rd and 4th boys on the cross country team. Rich improved his PR at the State championship course by over 60 seconds and Will improved by over 80 seconds! They have continued to show the results of their improved training this winter. Will, who did not advance to the Group 4 State Meet last spring in the 800m, won the Group 4 State meet in 1:59.2 and has run 1:57.8. Rich, whose PR from previous spring was 2:03, took 3rd in Group 4 State meet at 2:00.4 and ran 1:59 this season. They are more fit and as a result, more confident. Thanks, Shannon."

- Steve Shaklee- Head Coach Cherokee High School

Chapter 4- Physiological Profile Testing

"Shannon's methods are one of the most sophisticated

approaches we have seen in the United States."

- Jerry Cosgrove: Owner of Sports Resource Group and

Lactate.com - www.lactate.com

The "What," "Why," and "How" Of Human Performance

Physiological Profile Testing (PPT) is a method and testing protocol I developed to provide a view as to the "What," "Why," and "How" of human performance. It reveals "What" type of bioenergy an athlete has available to perform work, "Why" the athlete may or may not be improving in her/his training or performances, and "How" the athlete should steer her/his training to enhance physiology for better performances

PPT objectively assesses the "what" by measuring the full spectrum of bioenergetic output from aerobic to anaerobic via the use of net lactate, heart rate, velocity, power, capacity, and sport-specific exercise protocols. The PPT assessment goes beyond using the traditional method of "lactate threshold" and uses lactate dynamics to indicate the overall functionality and ability of each athlete to recruit required energy for their sport or event demands. PPT is indicative of an athlete's physiological readiness, response to training stress, and ability to adapt to training stress, which ultimately leads to optimal performance outcomes. The use of PPT data decreases trial and error training and response periods along with identifying if performance declines are due to physiological factors such as acute or chronic glycogen depletion, insufficient recovery, or inadequate training loads. The lactate dynamics across the human spectrum are indicative of the causes of performance declines but also enable coaches/trainers to address the "how" of training most effectively.

PPT objectively measures the "what" by quantifying eight Systems or areas of bioenergetic output. Similar to an engine's unique qualities of power versus efficiency outputs, each System has unique attributes of aerobic and anaerobic energy contribution. Each System is also unique in its ability to perform and adaptability to training. Understanding the basics of each System is of utmost importance because they are all critical components for enhancing overall performance outputs in any sport or event. PPT data will provide a clear picture of an athlete's current bioenergetic status so coaches can avoid unproductive or disadvantageous training.

PPT objectively quantifies the "why" by providing the current physiological readiness and bioenergetic availability of each athlete so coaches can prescribe the ideal training stimulus that each athlete needs for optimal performance. Given the fact that the current physiological blueprint of each athlete is most likely different, prescribing the same training stimulus for two different athletes most certainly will yield two different performance results.

PPT analysis will also indicate whether an athlete has a Bioenergetic Deficit. Bioenergetic Deficit means that the energy output, namely training load, is regularly exceeding energy input, rest, and fuel intake. When energy output is higher than energy input for more than a period of seven days, the ability to adapt to the current training load will be diminished. The decline of performance and health will be accelerated when the amount of deficit within one or more bioenergetic systems increases, as stated by the First Law of Thermodynamics, energy can neither be created nor destroyed but can change its state. If one has a bioenergetic deficit or the human system is lacking essential fuel or energy, there is no way for one to generate high work outputs. Adequate bioenergetic power is key to producing optimal performance results. Bioenergetic Deficit is manifested as stagnation or decline in performance or training. Via PPT data, daily training prescription is in a manner that is appropriate for an athlete's current physiological state so the athlete can regain functioning, bioenergetic availability, and improve performances.

PPT objectively addresses the "how" to improve bioenergetic availability and, ultimately, the performance by providing individualized heart rate data and training parameters (heart rate, pace, time, duration, volume, etc.) to achieve intended physiological responses within targeted Systems during each training session. By explicitly addressing each bioenergetic

System via the appropriate stimulus, the possibility for making intended physiological responses is considerably increased. The Second Law of Thermodynamics is the law of increasing entropy or level of disorder, randomness, or chaos within a system. This law also applies to human biochemical responses, the higher the randomness of the human system, the higher its entropy or disorder. Consequently, If arbitrary training stimuli are applied, randomized results will occur, which coaches will often blame on lack of fitness or mental fortitude when the reason may very well be a poorly organized system of stimuli. The more organized a system, the lower its entropy. If a systematic training stimulus is applied, then planned and predictable performance results will occur.

When To Use Physiological Profile Testing

PPT can be used at any stage of fitness or at any point during the year. PPT results show current physiological status to provide pertinent data to assist coaches in planning individualized training. Training phases are developed based on the physiological readiness of each athlete for an appropriate training load. Let me provide and all too common example.

A group of high school athletes arrives at the first pre-season cross country practice in late August. Some runners ran over the summer, but others decided to spend their time doing other activities. If the coach trained all those athletes the same because it was the beginning of the season, most of the athletes would respond sub optimally to the training stimulus. This would be because their physiological readiness and System strengths for certain types of workouts would be insufficient to sustain the workload.

Appropriate System targeting is crucial to athlete development and performance improvements. Attempting to train a System that is currently unavailable or before other Systems have achieved necessary rates or capacities will undoubtedly cause an athlete to fall short of his or her goal. Athletes who train at workloads that they are unprepared to handle will most likely get injured or stagnate in their performances. Ideally, PPT should be used every ten to twelve weeks, so appropriate adjustments may be provided for individualizing the training parameters for each athlete.

Why Use Physiological Profile Testing

Many coaches argue that they can use race performances or workout results to determine how an athlete is responding to training just as effectively, or more than tests of biological function. Yes, using performance and workout improvements are a simple, indirect way to assess physiological adaptations. If an athlete improves, one can assume the training stimulus is working. However, when an athlete plateaus or performances start to deteriorate throughout the season, the cause or causes for their poor performances can be much more ambiguous. Unfortunately, many coaches are quick to blame the athlete or say it is "mental," especially if other athletes on the team are performing well with the same training. It is more often the case, however, that the causes reside in a poorly systematized training plan.

PPT is a direct indicator of an athlete's current physiological status. As discussed earlier, using standardized percentages of maximum VO2, VO2 field testing, FTP, percentage of max heart rate, or "lactate threshold" will miss the mark for most athletes when trying to optimize performance. In such cases, Coaches and athletes would be better off making training decisions based on daily observation and educated trial and error.

The trial and error method is common practice in coaching. Unfortunately, it can take years for a coach to understand what type of training will work best for each athlete Because of the dynamic nature of human physiology, when a coach figures out what "worked" for a particular athlete that same stimulus will be unlikely to produce the same performance results the next time around. Using PPT data, a coach can quantify if an athlete is positively or negatively responding to the training stimulus as well as knowing whether the decline is mental or physiological. Objective, science-based data makes decision-making and steps towards physiological progress much more manageable.

Although the analysis of PPT data takes a considerable number of case studies to comprehend completely, it's important to note that a decline in physiological functioning or bioenergetic deficit is a red flag. The factors that can cause reductions in bioenergetic availability are usually: inappropriate or over training stimuli, poor nutrition, poor recovery, inadec hormone levels, metabolic pathway limitations, subop blood chemistry levels (iron, ferritin, Hct, Hb, RBC), viruses, disease, and sleep, although the causes of poor performance cannot be restricted only to this list.

Training stress is recognized as one of the most challenging stressors for human homeostatic feedback systems, Among the various symptoms of training stress are hormone imbalance, fatigue, abnormal biomarker levels, sleeplessness, among others. When athletes are showing signs or symptoms of inability to maintain healthy homeostasis, it is always better to address the cause first and foremost by prescribing appropriate training stress, instead of wasting an entire season or year by trying to treat or manage symptoms with supplements, hormone therapy, or foods. If an athlete's training load is beyond what he or she can handle given their current physiological state, the ability for the body to maintain homeostasis post-workout will become increasingly more challenging, leading to underperformance and certain of the other symptoms mentioned above. When a coach knows the adequate training load capacity for each athlete, performance improvements will be expeditious.

The PPT testing protocol involves ideally six to eight stages of two to four minutes of incrementally increasing intensity, either continuous or with less than 30 seconds rest between steps. On the last stage, athletes are asked to perform a maximal effort that is a shorter time or distance than the previous stages.

Although this is not an accurate indication of one's true maximum net lactate reading, that consideration is part of the PPT analysis algorithms. Net lactate samples are collected between each stage, along with heart rate, velocity, and power. PPT protocols should be sport-specific when conducted on an individual or team sport athlete.

In a team sports scenario, the goal is to monitor each individual and address the individual's needs. This should preferably take place in the off-season, to ensure the team unit has bio-energetically balanced individuals that can adapt to the team training load when the season begins. PPT can provide a means to effectively manage individual needs in team scenarios without disruption to team dynamics. PPT testing enables coaches to understand individual physiological differences versus a one-size-fits-all training prescription. PPT data virtually eliminates the guesswork in trying to figure out what "works" with each athlete. Following is a testimonial received from JJ Clark, who is now the Director of Track and Field at Stanford University. He also served a head coach for the 2008 U.S. Olympic Team. At the time of this testimonial he was head coach for the University of Tennessee women's track program.

In four years (2002-2005), the University of Tennessee Women's track program implemented physiological profile testing data to turn an ordinary track team into NCAA team champions (2002-NCAA 53rd place, 2003- 23rd place, 2004- 4th place, 2005- 1st place). The University of Tennessee Women's track program continued to dominate the NCAA during Clark's tenure while using physiological profile data.

"Since 1999, Shannon has provided pertinent testing and data that has helped in the development of optimal training programs for my athletes. Shannon is an expert in the application of Biochemistry and Physiology as well as lactate testing data for training and competition. Physiological Profile testing allows each athlete to get the most out of their abilities."

-JJ Clark: Director of Track & Field- Stanford University and 2008 Olympic Team Coach

8 part test –
4 parts for cults at least
< 5 parts showing bad energetic deficit
7-10 days of glycogen sparing

tapering - gain a system, but usually training depleted because you're training depleted

3 or less = carb/energy issue

Chapter 5. Bridging The Gap - Data And Performance

Bioenergetic Systems

A typical Physiological Profile test results in scores that indicate the potential of eight bioenergetic Systems in every human. These eight Systems are measured via net lactate, velocity, power, and heart rate data. By doing so, a PPT identifies an individual's energy availability across the entire human bioenergetic spectrum. The eight Systems all have unique characteristics that include a percentage of aerobic or anaerobic contributions, trainability, power output, velocity output, and capacity limits. Each System is measured, quantified, and addressed explicitly according to each person's current power level.

Even though Systems can be vastly different in objectives, they are all equally important to human performance outcomes. Therefore, it is imperative to understand the roles and goals of the Systems as they relate to various performance tasks. The following is an overview of each of the eight bioenergetic Systems and their physiological objectives.

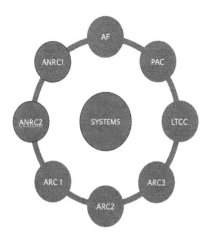

Sample System Based Training PPT Report

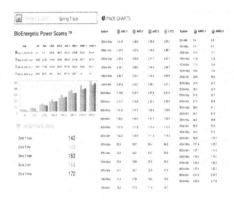

System: Aerobic Foundation ™(AF)

Ability to perform work using primarily aerobic energy sources such as free fatty acids and blood glucose. Aerobic foundation training can range from five minutes to hours of continuous low-intensity work performed at Zone 1 to Zone 2 heart rates.

Physiological Objective: To develop the foundation of the aerobic system and increase the following key aspects: fat metabolism, the number of aerobic enzymes, the size and number of mitochondria, capillarization, and lactate clearing rate.

System: Prolonged Aerobic Capacity™ (PAC)

Ability to perform work using primarily aerobic energy sources such as free fatty acids and blood glucose. Prolonged Aerobic Capacity training totals 60 minutes or more of continuous low-intensity work performed at Zone 1 to Zone 2 heart rates. Physiological Objective: To develop maximum speed at which one can train for 60 minutes or longer.

System: Lactate Tolerance, Clearing, and Capacity™ (LTCC)

Ability to perform work using primarily aerobic energy sources such as blood glucose along with glycogen. Lactate Tolerance, Clearing, and Capacity training totals up to 60 minutes or more of continuous work or 5 minutes (2 minutes if using LTCC pace) and longer intervals with less than 2 minutes rest at moderate intensity performed at Zone 3 heart rates or LTCC pace. Physiological Objective: Increase lactate clearing, lactate tolerance, and sub-maximal aerobic capacity.

System: Aerobic Rate Capacity™ 1, 2 and 3 (ARC1, ARC2, ARC3)

Ability to perform work using primarily aerobic energy sources such as blood glucose along with glycogen. Aerobic Rate Capacity training totals up to 45 minutes of 1 to 10 minutes of work intervals with 1-5 minutes rest at moderate intensity performed at specified ARC System pace.

Physiological Objective: To develop and increase major training adaptations: stroke volume, maximal aerobic capacity, maximal aerobic rates, the velocity of maximal rate of oxygen consumption, and lactate production.

System: Anaerobic Rate Capacity™ 1 and 2 (ANRC1, ANRC2)
Ability to perform work using primarily anaerobic energy sources such as glycogen. Anaerobic Rate Capacity training totals up to 15 minutes of 10 seconds to 3 minutes work intervals with 1-12 minutes rest at high intensity performed at specified ANRC System pace.

Physiological Objective: To increase maximum lactate production, maximum anaerobic capacity, lactate buffering capacity, and rate of anaerobic respiration.

Training Systems

In addition to the eight bioenergetic Systems, there are three accompanying training Systems applied to precisely achieve all possible physiological and performance objectives in the human realm. The three training Systems fall within one or more of the bioenergetic systems. Still, the parameters in which to execute their specific physiological objective are different than another comparable bioenergetic system. The use of any of the 11 Systems can target physiological and performance goals, but they all differ in terms of bioenergy, trainability, velocity, power, and capacity. The following is an overview of the three new training Systems and their physiological objectives.

System: Recovery and Maintenance (R&M)

Ability to perform work using primarily aerobic energy sources such as free fatty acids and blood glucose. Recovery and Maintenance training is less and 45 minutes of continuous low-intensity work performed at Zone 1 to Zone 2 low heart rates. Physiological Objective: To promote recovery, gluconeogenesis, and lactic acid removal following glycogen-depleting training such as high-intensity intervals or workouts longer than 60 continuous minutes. Maintenance of cardiovascular and skeletal muscle adaptations.

System: Neuromuscular Adaptation Glycogen Sparing™ (NAGS)

Ability to perform work using primarily aerobic energy sources such as free fatty acids and blood glucose. Neuromuscular Adaptation Glycogen Sparing training should be less than 30 minutes in length and it should consist of 20 to 60 seconds of interval work with 1-2 minutes rest between each interval at specified System pace.

Physiological Objective: To increases neuromuscular economy, efficiency, speed, muscular/tendon elasticity, and strength without glycogen depleting effects.

System: Hybrid MCT™ (HMCT)

Ability to perform sustained velocity or power in events lasting less than 10 minutes. HMCT training is 5-10 minutes of ANRC1 or ANRC2 work followed by 10-30 minutes of ARC1, ARC2, or ARC3 work.

Physiological Objective: To increase the stimulation of monocarboxylate transporters and lactate shuttling efficiency at Aerobic Rate Capacity™.

The following is another testimonial, this time from Ryan Comstock a highly successful track coach.

"I started using Shannon's physiological profile testing after 11 years of coaching and developing 11 state champions. I felt the need to do something different and better. Having already known Shannon for many years, I first witnessed her testing when I helped pace my former athlete, Liz Costello, through a testing session. Liz's 5k and 10k times saw huge improvements under Coach JJ Clark and Shannon's testing, so I was very interested in the scientific individualization. Getting a few of my top athletes to test for the first time was an exciting new experience. It was such a great feeling to know exactly what each individual needed to work on and how. Come pre-season, I no longer needed to use a time trial or a race to determine who ran over the summer. Not only do I now know who ran, but more importantly, I know who ran the smartest. After several years I now have the majority of my distance runners testing four times per year. I see more and more athletes running faster with greater ease. My athletes are now doing workouts based on their physiological readiness and not just based on a calendar. I give my highest recommendation of Shannon's physiological profile testing for any athlete or coach who is looking to develop fast and healthy athletes."

-*Ryan Comstock - West Chester University- Head Cross Country & Track Coach*

Heart Rate Based Systems (R&M, AF, PAC, LTCC)

Heart rate monitors have been in use by athletes for more than 30 years and is the most common biometric used during training. Despite its popularity, many coaches, athletes, and fitness trainers are still using heart rate data ineffectively for maximizing performance and weight loss. However, as one expert has stated,

"Despite 61 million wearable fitness devices being sold worldwide last year, new data has found that the majority of people working out are still not fully clear on how to use a heart rate monitor to maximize their fitness goals. The research from Polar, a creator of wearable sports and fitness technology, has revealed that athletes, regardless of fitness level, are, on the whole, only passively tracking their heart rate rather than actively applying it to workout regimes."
- Dave Claxton "SportTechie", October 2017.

Throughout my years of experience working with a myriad of coaches, athletes, and fitness enthusiasts, the most common thing I stress is the importance of the "HOW" and "WHEN" to use heart rate in training. The use of heart rate is ONLY effective when used correctly.

It is critical to understand why heart rate is a biometric that is used to indicate how hard the body is working and why heart rate varies during exercise to understand the basic concepts of its use. Heart rate is one of the variables that contribute to cardiac output. Cardiac output (CO) is the number of times the heart beats per minute, heart rate (HR), multiplied by the amount of blood pumped with each beat, stroke volume (SV), or $CO = HR \times SV$. Cardiac output is different for everyone and largely depends on a person's size, as an adult heart usually pumps about three to four liters per minute of blood at rest. During exercise, the demand for oxygen increases throughout the working muscles, and the heart will increase the amount pumped per minute to about nine to 12 liters per minute. The factors that affect heart rate are oxygen demands, autonomic nervous system function, hormones, fitness levels, and age. The factors that affect stroke volume are heart size, fitness levels, gender, ejection fraction, contractility of the left ventricle, duration of the contraction, preload or end-diastolic volume, and afterload or resistance. Unfortunately, measuring cardiac output is quite an invasive procedure in which devices such as a pulmonary artery catheter, an echocardiogram, or an arterial pulse waveform analysis.

Heart rate is the one variable of cardiac output that is easily measured and can give some indication of increases in oxygen demands required by the body during work. During an exercise session, the heart rate is also one of only two variables that contributes to cardiac output that can vary greatly while also being the easiest one of the two to measure. Heart rate can vary between as low as 36 to as high as 220 beats per minute, a factor of 6, while stroke volume can vary between 70 and 120 milliliters, a factor of only 1.7. Due to the high factor of the variability of heart rate in indicating increased demands of the body, the use of heart rate as an indicator of exercise intensity is most effective when the heart rate remains relatively constant for five minutes or more.

Keys to using heart rate as a training parameter:

1. Use heart rate during low intensity, continuous efforts, in training zones 1 or 2 continuously for five or more minutes with less than one-minute rest.

2. Use of solely heart rate to achieve training objective is valid when targeting the R&M, AF, and PAC Systems.

3. Use heart rate or pace during moderate-intensity intermittent intervals longer than five minutes with less than two-minute rest periods. If done at LTCC heart rate, zone 3, the use of heart rate can be effective for targeting the LTCC System.

4. Refrain from using heart rate during intermittent, high-intensity interval training. Heart rate is ineffective in targeting ARC3, ARC2, ARC1, ANRC2, or ANRC1 Systems.

5. Resting heart rate is expected to vary daily. Adjusting the training load daily based on an acute

66

resting heart rate variability is not recommended for effective training responses.

Only by testing will one know their actual individual heart rates and heart rate training zones. Many are unaware that heart rate ranges and zones can fluctuate as one experiences Bioenergetic Deficit or other chronic physiological changes. So, testing should be conducted at least every six months to ensure proper heart rate zones are being used in training. The popular method of using heart rate formulas based on 220 minus age is not accurate enough for most active people, especially athletes. Having accurate information and appropriate heart zones can be obtained via physiological profile and max heart rate testing. Incremental maximum exercise tests in various modes such as running, biking, swimming, or rowing are used to calculate current heart rate zones.

HR variations

Heart rate is only an accurate indicator during low intensity, continuous effort training sessions because the daily, fluctuations in energy system contributions for the R&M, AF, and PAC aerobic energy systems are highly volatile. Essentially, the pace or speed one needs to go to keep the body using low bio-energy can be influenced by acute factors such as recent hard efforts, inadequate recovery, dehydration, heat, humidity, or lack of fuel. Use of heart rate zones governors will make sure one is working at the appropriate velocity or power output for that day, when one wishes to train in the lower bio-energy zones. Using pace or power targets for continuous aerobic efforts or recovery days will often lead to over-working, inadequate recovery, and ineffective low-end aerobic energy system stimulation.

For example, let's take a cyclist with a low-end aerobic heart rate zone target of 124-140 beats per minute. She rides for 60 minutes at 124-140 beats per minute, with an average power output of 170 Watts. Two days later, after a volume-intensive and high-intensity interval session, she rides for 60 minutes at 124-140 beats per minute, with an average power output of only 150 Watts. Why the difference? Shouldn't she try to ride the same power target for all her recovery or aerobic rides? Shouldn't she push through and ride her "easy" Watts, no matter what? The answer is no, quite the contrary. One may think if the heart rate is influenced by environmental factors such as heat and humidity or by nutritional factors such as dehydration, how can it be indicative of aerobic energy contribution? These factors are precisely why heart rate monitors work optimally for low-intensity continuous efforts. Heart rate can indicate how one is coping with environmental conditions or daily stress and how hard the body is actually working, even if the velocity or power output is slower than average.

Be aware that using pace, power, RPE, or "feel" to determine training efforts for R&M, AF, or PAC Systems will frequently lead to Bioenergetic Deficit, performance plateaus, performance decreases, and inefficient metabolic output. The beauty of using heart rate in this manner is that it is impartial to feelings or perception. Although some days the prescribed zone one or two heart rates may feel "too slow" or "too easy," that is what the body needs on that particular day.

Adhering to proper heart rate zones will make more significant performance gains and increase metabolic efficiency across all Systems. Going slow or easy some days will enhance one's ability to go hard on others. Improving low-end aerobic energy as well as high-end intensive energy will make the body efficient at burning fats and carbs during exercise and also at rest.

"I started working with Shannon in the late Fall of 2017. I was able to use three of my athletes who have been running and jumping at high levels and were upperclassmen. What started as a curiosity of having them test and saw the results ended up making me adjust my approach to specific ways of training. My two best examples are first, Athlete "A," a 60-meter specialist who also does triple jump. As a shorter, very stocky athlete, we needed him to be able to drop healthy weight and increase his fitness. To start, his warm-up, with the prescribed heart rate based on his testing, was monitored by a heart rate system we use. We found something as simple as changing his warm-up duration time and what target heart rate zone to stay in rapidly increased his fitness at the most basic level. It was hard for him at first, but each week it got more comfortable for him. His workout plan was adjusted, and in a month and a half, his weight was back to where it should be, his fitness level increased, and his performances improved. I had him wear a heart rate monitor for a meet where he took six jumps and had trials and finals in the 60m dash. I recorded all the data. I

repeated it a month later, where the format was the same. His recovery in between the jumps and rounds of the 60m was quicker, and he was performing better. That lack of recovery was a significant problem for him in the past.

Athletes "B" and "C," both 400m runners with similar times, tested as well. "B" was shown to have a high level of Aerobic Capacity but was lacking in Anaerobic. "C" showed high levels of Anaerobic capacity, but lacking in Aerobic. Again, different heart rate targets for the warm-up and the science-based workout times and recovery, although slightly different, allowed both to continue their strengths and increase the energy systems they were lacking. When they tested again with Shannon, the levels that were lower for each "A," "B," and "C" were significantly higher, and other systems were now catching up as well. It's incredible how adhering to System Based Training principles did for my athletes."

-Brian Hirshblond- Monmouth University, Associate Head Track & Field Coach

Velocity And Power Based Systems (LTCC, ARC3, ARC2, ARC1, ANRC2, ANRC1)

Velocity or power targets are the most effective means to indicate intensity when targeting ARC and ANRC Systems, which involve performing intermittent, higher intensity interval training. As stated previously, due to the variability of heart rate for indicating increased demands of the body, the use of heart rate as an indicator of exercise intensity is most effective when it remains relatively constant for five minutes or more. Therefore, using heart rate during intermittent or interval training will give an inaccurate indication of the intensity and is unable to target specific energy system stimulation. Precise velocities, power outputs, along with appropriate work volume and work to rest ratio, are the best way to target ARC and ANRC System stimulation. The LTCC System can also be targeted via pace or power if performing intervals longer than 2 minutes with less than two minutes rest at LTCC intensity. During intermittent and high-intensity interval training, an individual's physiological makeup and the amount of work and rest will skew heart rate values in such a way that using heart rate as a governor for this type of training will fail to elicit a specific physiological response. Repeated use of heart rate for high-intensity interval training will increase the risk of injuries, negatively impact the metabolic system, and can cause metabolic inefficiency or weight gain.

Heart rate response is also an unreliable metric for use in indicating increased oxygen demands, lactate, and blood flow during high intensity intermittent training due to the num variability of lactate dynamics that occurs during such training. Athletes will generally find themselves unable to achieve the intended physiological objective by using heart rate as their guide to elicit a performance response during high intensity,

71

intermittent training, and, therefore, will be unable to achieve the intended physiological objective. Proper stimulation of all bioenergy Systems increases metabolic and metabolic and performance outputs. When using heart rate as a target for high-intensity intervals, Injuries, metabolic inefficiency, Bioenergetic Deficit, and performance declines are often the result of this approach.

"Shannon and her team at Go! Athletics helped me prepare for CCCAN 2015! I won two silvers and two bronze medals posting best or near best times in all 5 of my events (50m Breast, 100m breast, 200m breast, 200m IM, 50m free). Her approach focused on my specific events and addressed my areas of weaknesses. I feel great every time I get in the water and can't wait to race again! Thank you, Shannon."

-Maddie Pujadas- Columbia University Swimmer

System Training Prescription

R&M- Recommended parameters are as follows
Session Volume: Less than 45 minutes
Sessions Per Week: one or more
Target Heart Rate: Zone 1 to 2 low
Enhances lactate elimination or clearing

AF- Recommended parameters are as follows
Session Volume: 30 minutes or more

Sessions Per Week: one or more

Target Heart Rate: Zone 2

Enhances lactate elimination or clearing

GROUP TRAINING CALENDAR BY PHASE : GROUP M.KOLOR

GO/ATHLETICS

System AF (06 October 2019 - 16 November 2019)

PAC- Recommended parameters are as follows

Session Volume: Up to 90 minutes

Sessions Per Week: one or more

Target Heart Rate: Zone 2 mid to high

Interval Duration: 6 minutes or more

Interval Rest: Two to three minutes

Pace: PAC

Enhances lactate elimination or clearing

LTCC- Recommended parameters are as follows

Session Volume: Up to 60 minutes

Sessions Per Week: one to two

Target Heart Rate: Zone 3

Interval Duration: 2 to 30 minutes

Interval Rest: One to two minutes

Enhances lactate elimination or clearing

ARC3- Recommended parameters are as follows

Session Volume: Up to 45 minutes

Sessions Per Week: one to two

Target Heart Rate: N/A

Interval Duration: 3 to 10 minutes

Interval Rest: 1:1 up to 5 minutes

Enhances lactate production

ARC2- Recommended parameters are as follows

Session Volume: Up to 40 minutes

Sessions Per Week: one to two

Target Heart Rate: N/A

Interval Duration: 1 to 5 minutes

Interval Rest: 1:1 up to five minutes

Enhances lactate production

ARC1- Recommended parameters are as follows

Session Volume: Up to 30 minutes

Sessions Per Week: one to two

Target Heart Rate: N/A

Interval Duration: 1 to 3 minutes

Interval Rest: 1:1

Enhances lactate production

HMCT- Recommended parameters are as follows

Session Volume: Up to 3 minutes ANRC and up to 30 minutes ARC

Sessions Per Week: one to two

Target Heart Rate: N/A

Interval Duration: see ANRC and ARC details

Interval Rest: see ANRC and ARC details

Enhances lactate shuttling

ANRC2- Recommended parameters are as follows

Session Volume: Up to 15 minutes

Sessions Per Week: one to two

Target Heart Rate: N/A

Interval Duration: 1 to 3 minutes

Interval Rest: 1:3 to 1:4

Enhances lactate production and buffering

ANRC1- Recommended parameters are as follows

Session Volume: Up to 5 minutes

Sessions Per Week: one to two

Target Heart Rate: N/A

Interval Duration: 10 to 60 seconds

Interval Rest: 1:3 to 1:4

Enhances lactate production and buffering

Chapter 6. Quantifying and Predicting Human Potential

Lactate is the human energy currency. BioEnergetic Power Scores (BEPS) quantify that currency and performance output ability.

"Shannon's expertise and knowledge helped me run my personal best in the 800m, win 7 US National titles, and make 3 Olympic teams. PPT and SBT data can help any athlete take their performance to the next level".

-Hazel Clark: 3 Time Olympian (Track & Field- 800m) Nike Athlete

The BioEnergetic Power Score (BEPS)

Training "harder" than ever and having poor performances? Doing more volume with less return? Epic workouts and worse performances? How do we, as athletes or coaches know when more or harder training stagnates performances? How can one measure or track how the body is responding to the training? Is all this hard work pounding the pavement, tearing up the track, logging massive volumes in the pool, or on a rowing machine worth all the time and energy? One of the unique aspects of System Based training has been the development of BioEnergetic Power Scores or BEPS. BEPS can precisely determine those age-old questions quickly, accurately, and effectively. BEPS give us the capability of quantifying power scores for each metabolic system with the same units so they can be compared with one another to determine their relative contribution to one's overall performance. BioEnergetic Power Scores quantify internal human power output, are applied to daily training plans, and gauge performance capabilities. BEPS provide a way to translate the power scores into the same "currency" so those scores can be compared , strengths and weaknesses noted, following which a training plan can be developed to improve those systems within the bioenergetic system that are weak, while maintaining those that are strong or adequate at an optimum level.

Just as engineers can design, measure, calculate, and assess the performance capabilities of engines, physiologists can do the same with human performance capabilities by measuring bioenergetic availability via sport-specific blood lactate protocols. Human bioenergetic power output can be measured precisely with a distinctive set of variables combining physiological, biochemical, and movement data resulting in a BioEnergetic Power Score (BEPS) for each System within the aerobic and anaerobic metabolic spectrum. Each activity or athletic event has a distinct set of measurable variables that correlate to the optimal bioenergetic power for that particular event. For example, the bioenergetic power required to run a 2 hour and 30-minute marathon is greater than that needed to run a 4-hour marathon.

Calculation of BEPS involves metrics from the Physiological Profile test such as net lactate, velocity, power, and work rates. BEPS scores are indicative of each System's constant net energy output. System BEPS ranges are unique to each System, and the value indicates at what level each System is operating. The combination of each System's BEPS values will all contribute to an athlete's ability to perform specific tasks as well as particular performance marks in measured variable sports such as running, swimming, and rowing. Comparative analysis of System BEPS scores from test to test can indicate which areas of physiology are improving or which areas require more development to achieve desired performance outcomes.

To accomplish any work task, whether it is a competitive event or life, a certain amount of bioenergetic power is necessary to complete those tasks. In sports performances, where the goal for an athlete is to maintain a particular velocity over a specified distance, BEPS information can indicate whether it is possible to achieve that performance time. Each performance time for any distance event requires a certain level of physiological development. BEPS will gauge what performance time an athlete is physiologically capable of producing. The BEPS metric is also a quantifiable and reliable indicator for sports scientists and coaches to be able to prescribe subsequent training phases or Systems focus for improved accuracy and peak performance results.

As an example of what can be achieved with System Based Training, below is the BEPS chart progression which displays the effect of that training on the bio-energetic profile of a male high school runner who was evaluated in March 2014. He was struggling to run 4:30 in the 1600m when he had previously run 4:15. After several physiological profile tests and System Based Training prescription, this athlete ran a 4:08 mile indoors later that year. More specific on this athlete's training prescriptions are provided in Chapter 10.

	AF	PAC	LTCC	ARC-3	ARC-2	ARC-1	ANRC-2	ANRC-1
Dec 13, 2014 01:01	47.02	73.22	135.54	227.78	308.76	395.42	487.26	644.99
Sep 10, 2014 01:01	48.18	78.7	150.26	225.92	299.33	398.78	526.76	0.0
Jul 09, 2014 01:01	43.76	76.07	142.54	216.7	302.5	386.28	456.52	0.0
Mar 08, 2014 01:01	40.57	74.95	163.68	248.17	0.0	0.0	0.0	0.0

Genetics is static, but Physiology is dynamic.

Human genetics will remain unchanged as one develops and trains, but one's physiology is ever-changing. Humans are living systems that produce ATP from organic sources such as carbohydrates via oxidative phosphorylation. Oxidative phosphorylation, or electron transport chain, is a highly efficient metabolic pathway where cells use enzymes to oxidize nutrients to produce ATP. Every activity from sleeping, hiking, rock climbing, rowing, running marathons, or swimming a 50m butterfly are defined along the human bioenergetic power continuum. Fortunately, human bioenergetic systems have a distinctive set of variables that can be quantified and measured by combining physiological, biochemical, and movement data. Each activity has a set of measurable variables that correlate to the optimal bioenergetic power required for that particular activity or performance mark. For example, for one athlete to run eight minutes for 1600 meters, it will require less bioenergetic power across all Systems than necessary to run four minutes for 1600 meters. See above 1600m reference chart for BEPS performance requirements.

Human power, fuel efficiency, performance capabilities, or different physiological makeup and energy systems are unique, similar to the way various car engines are unique in their qualities. When measuring an athlete's current bioenergetic availability, it is possible to determine if the athlete is capable of performing at various levels in a sport or event. A minimal amount of bioenergetic power must be available for athletes to achieve multiple performance levels. No amount of mental toughness will enable an athlete to accomplish a result or outperform their current bioenergetic availability. BEPS data are an essential piece of information used to create highly specific, individualized athlete training plans based on physiological data versus performance data. Engineers can measure a car engine's horsepower, fuel efficiency, and performance capabilities. BEPS metrics give the same insights for human performance as engineers measure for engine performance capabilities.

Let's compare two very different vehicles: the 2018 Dodge Challenger SRT Demon and the 2018 Mitsubishi Mirage. The Demon has 808 horsepower and is touted as the most "powerful production non-supercar ever built, an American car that is." The Demon is undoubtedly considered a high-performance vehicle with a 6.2 liter, eight-cylinder engine and can run on 100 octane gasoline to generate up to 840 horsepower. The Demon also has good acceleration speeds that are good for 0 to 60 miles per hour in a time of fewer than 2.5 seconds. Also, the Demon will go a quarter-mile in less than 10 seconds but is extremely fuel-inefficient at a whopping 16 miles per gallon of gas. Compare the Demon to the Mirage which is touted as "the least powerful car available today" with a 1.2 liter, three-cylinder engine providing just 78 horsepower. The Mirage has slow acceleration speeds taking 13.4 seconds to go from 0 to 60 miles per hour. The Mirage is certainly doubtful as a high-performance vehicle. However, the Mirage's unimpressive power is often preferred by consumers due to its impressive fuel economy. The Mirage's most redeeming quality, especially for a non-electrified vehicle, is being one of the most fuel-efficient cars available with a rating of 39 miles per gallon of gas.

Each of the two engines is designed with features that achieve different performance objectives as some consumers love horsepower, and some love fuel efficiency. Each performance objective makes something different and desirable, but there is always a give and take with energy output. For example, the high performing engine with monster horsepower burns fuel very quickly, but the low performing engine with little horsepower has high fuel conservation. The energy of each engine has been designed to perform at various output levels based on principles of engineering and thermodynamics. When an engine is underperforming, a mechanic can run diagnostic assessments on the engine and fix the problems. System analysis and BEPS calculations provide a previously unexplored perspective into individual performance dynamics. The BEPS are an essential piece of information in the System Based Training methodology. A fundamental component of SBT's dynamic periodization positive response stimulus model is the continuous use of current PPT and BEPS data to create highly focused and individualized athlete training parameters in every System.

Ongoing testing provides measurable and quantitative values in each System that also directly correlate to performance. Some of the problems of evaluating training effectiveness only on performance changes as compared to evaluating it with quantitative data that is based on bio-energetic power are as follows.

1) performance change is unable to identify which Systems or areas need physiological adaptation and

2) BEPS allow coaches to understand if the training stress needs to be modified due to Bioenergetic Deficit. BEPS, over time, show individual responses to various types of training stimuli and enable coaches to eliminate many months or even years of trial and error.

The BEPS quantifies the bioenergetic power in eight Systems: AF, PAC, LTCC, ARC-3, ARC-2, ARC-1, ANRC-2, and ANRC-1. The BEPS values in each System will indicate whether that System is currently available to adapt to targeted training stress and at what level that particular System is operating.

BEPS values give insight as to the bioenergetic availability of internal fuel sources and one's current physiological ability to produce required outputs for performance purposes. Coaches and athletes can achieve great insight as to the actual physiological effects of the training load and intensity given to their athletes and how it can influence performances in various events through the use of BEPS. Coaches and athletes can use BEPS data to modify or steer training periodization or focus on peak performance results. Implementing regular PPT and BEPS data into daily training ultimately gives coaches direct and reliable measures to eliminate the guesswork in developing their training plans.

BEPS levels in each System have a direct correlation to an athlete's ability to achieve specific performance marks in all measured variable athletic events. For example, if the BEPS score is too low in the Systems relevant to produce a 6:00

minute 2000-meter rowing score, the athlete will be physiologically unable to achieve that performance goal. The athlete needs BEPS in the Systems that will enable the goal for the 2000-meter time to be met. The training stimulus can then be targeted to improve the BEPS in the System or Systems necessary for various performance goals, instead of guessing what type of training each athlete needs to meet their goals, BEPS allows coaches to map a precise, individualized training periodization/plan.

Let me cite an example that illustrates the importance of accurate quantification in the sport of rowing.

"Rowers feel that they are always subject to the coach's evaluation and scrutiny. They row hard away from the dock, and they don't stop until they return. The typical Steady State session ultimately turns in to a competitive row. Therefore, the majority of training is either all out or one that is undoubtedly not in the realm of recovery or aerobic foundation. There is a mindset that if something worked for one school, and that school was successful, then that training plan becomes the one that is followed. Training plans are shared, recycled, and maybe slightly modified from coach to coach, program to program, but most are fundamentally similar. Typically, coaches were rowers who followed what their coaches prescribed for them without much questioning or skepticism.

Will Purman came to me as one of those athletes who had spent years going hard all the time and faithfully doing what he was told. He is competitive by nature and did what his coaches asked of him, and he was successful at the collegiate level. As he had begun to, He finally became open to an alternative approach to training when he embarked on an elite path, and one where there wasn't a legacy foundation in place."

Christopher McElroy- Elite Rowing Coach

Will's BEPS Progression

Typical rowing analysis/training recommendations are concerned primarily with intensities described as "Steady State," "Tempo," and "Threshold" efforts. These types of efforts are associated with energy contribution and physiological adaptation in the first three Systems; AF, PAC, and LTCC. In the example BEPS chart above, Puerto Rico Men's Rowing National Team athlete, Will Purman, displayed a typical rower's physiological profile upon initial evaluation with the more aerobically based Systems available for contribution to performance. To most scientists or coaches, it would appear from the initial assessment that Will is more of a slow-twitch, Type I fiber type athlete due to his low max blood lactate values and the unavailability of his ARC1,

86

ANRC2, and ANC1 Systems. What was deficient were contributions from the ARC1 and ANRC2 Systems for the duration and intensity of an event like the 2000-meter race. Our testing showed that adequate contributions from these two Systems were needed to generate sufficient power and velocity in order for him to improve his performance in the 2000-meter event.

After his initial PPT in December 2013, Will followed SBT periodization and training recommendations for 12 weeks and was re-tested in March 2014. Will also performed a 2000-meter erg time trial at 6:36 the week before his December 2013 PPT. The intent with Will's training stimulus was to maintain power and increase his capacity in currently available Systems (AF, PAC, LTCC, and ARC3) and activate his ARC1 and ANRC2 Systems. As anticipated, Will's March 2014 PPT displayed that his training adaptation resulted in maintained power, increased training capacity in his AF, PAC, LTCC and ARC3 Systems, and activation of ARC1 and ANRC2. Subsequently, Will also performed a 2000-meter erg time trial at 6:19 the week before his March 2014 PPT. Following is a testimonial from Will regarding his experience with SBT.

"After taking an extended break from rowing in 2015, I was given the rare opportunity to represent my second nationality (Puerto Rico) at the world rowing championships. With only about eight weeks to go before the World Championships, I rushed to Shannon for her expert advice. With Shannon's personalized System Based Training plan, I was able to get back to international level speed in a very short amount of time and *set a flat water PR. I look forward to continuing System Based Training in my preparation for Olympic qualification."*

-Will Purman- Puerto Rico National Team- 2015 World Championships qualifier

Bioenergetic power in each System will change throughout different training phases and understanding the relevance of each power score for the event is fundamental to improved performances. The analysis of bioenergetic power in each System is a valuable tool for understanding current performances and physiological functioning. Bioenergetic power analysis is a crucial variable to quantify training progress and to provide a focused training stimulus that will lead to optimized and accelerated performance results. Training recommendations are individualized based on what bioenergetic outputs each athlete can handle at any given time. BEPS data reduces the risk of injury due to overstressed or overreaching training.

Interpreting BEPS Charts

"The human body is one big math equation." -Mark Cuban

Nearly all athletes experience times when they have great workouts, but performances just aren't matching up. The coach and athlete start to question what the cause could be. Still, after much deliberation, solutions typically come down to a guessing game of altering training, trial and error workouts, or even mental exercises until the performances start improving. This approach may take weeks, months, or even years to have success if success comes at all. The quantification of BEPS throughout the various systems can take the guesswork out of this process and greatly increase the probability of good performances.

BEPS is a metric derived from a culmination of an eminence database. This database evolved while conducting long-term, extensive field testing, and analysis in collaboration with multiple elite coaches, thousands of athletes, and several peer physiologists. The efficacy of training effects derived from the SBT approach were evaluated longitudinally from 1999-2014 amongst a heterogeneous athlete population. Over 90,000 samples from recurrent athlete assessments of four to six tests per year over two to six-year period were included in the analysis. Physiological data was compared to corresponding performance data. By using comprehensive physiology evaluations, the SBT model of sophisticated sports performance monitoring and assessments, has been highly effective in facilitating improved performances and physiology.

A thorough analysis of biochemical and physiological data enables us to give coaches and athletes accurate performance targets (within measured variable sports/events), with over 95% of the athletes tested. Coaches and athletes can honestly assess current performance expectations as well as understand which

89

areas of training to target to achieve a goal or improve current performance levels.

Validation of The Use of BEPS in the SBT Methodology

BioEnergetic Power Scores illustrate individual System responses to various training stimuli and enable coaches to gain valuable insight to develop highly specific individualized training plans. BEPS also measures at what level an athlete can currently perform for different distances, events, and sports. By archiving individual athlete performances, from Olympians to weekend warriors, a data set was developed that systematizes the BEPS levels required for specific performances across various events. The BEPS target performance charts allow coaches or athletes to compare BEPS data to the targeted BEPS for different performance ranges.

The following is a prime example of how dynamic physiology, along with the appropriate stimulus, can yield highly predictable physiological responses. The following is a case of a professional female marathon runner, Liz Costello. Liz was experiencing performance declines, inability to recover, nagging injuries, and a deteriorating desire to continue training or racing. Liz performed three Physiological Profile tests over eight weeks. Her BEPS data objectively quantified her current performance capabilities as well as the appropriate training stimulus to create increases in her bioenergetic availability and power. She described her experience with our SBT approach as follows.

"In the fall of 2017, my coach and I were not sure if I needed more or less training to improve on poor workout results. I had taken seemingly adequate periods of rest or "downtime" during the summer, and there was no shortage of emotional or mental motivation in the fall. However, I struggled to hit typical longer interval workout times without becoming lactic or to maintain what would be my usual pace on threshold runs. I began to doubt if I was trying hard enough or if my iron levels were in range. Based on years of training and lab results, I established that neither of these possibilities explained my current state. Having worked with Shannon Grady in the past, first as an athlete at the University of Tennessee, I gave her a call and asked her opinion on the question my coach and I were facing, "Do I train more or less right now to improve performance?" Shannon explained that the answer was essentially neither, but that my training needed to be adjusted based on where my body was at physiologically. She explained that the symptoms I described were familiar, but the specifics on how to proceed would require that I complete a physiological profile test. Upon receiving the results and analysis of that test, I felt a tremendous sense of validation and clarity because not only did the data provide an answer - fatigue - but it also provided a way forward. From this point, Shannon's work and consultation over a few additional tests helped to provide a systematic and measured approach to helping my coach, and I implement training that restored my full range of physiological

ability. And as results met expectations with each test, my enthusiasm for the upcoming season also increased."

- Liz Costello- Professional Runner

The following is an overview of Liz's PPT testing analysis, results, and recommendations from December 2017, January 2018, and February 2018. Having tested Liz in college, I knew she was only using 25-30% of her bioenergetic potential with only 2 Systems currently available. My previous evaluation of her December 2017 PPT showed that she typically had 7-8 Systems available. I was aware that her training focus from 2015 to 2017 had shifted from 3-10-km racing to half and full marathon racing, so there was going to be a natural shift to the left or aerobic side of the bioenergetic spectrum from her collegiate testing results. No matter what event an athlete is preparing for, performances are most efficient with at least 6 Systems available from either the right or left side of the bioenergetic spectrum. At a minimum, the goal I decided upon was to ensure that Liz had the six aerobically dominate systems (AF, PAC, LTCC, ARC3, ARC2, and ARC1) available at any point in her training phases, especially for marathon preparation. Even though she does have the genetic potential to access all eight Systems she was advised that the availability of her anaerobic systems was unnecessary for optimal marathon preparation.

In the past, Liz had achieved the BEPS requirements to perform a sub 2:30 marathon, so getting back her physiological and bioenergetic potential was a question of how to get her there, not can we get her there. Her goal was to run a sub 2:30-marathon in April of 2018. Liz's genetic potential was such that I concluded that she had plenty of time to accomplish her current marathon goals. We would achieve this by regaining bioenergetic availability, power, efficiency, and then finally building capacity in various areas pertinent to marathon racing.

Below are the BEPS targets for a 2:30 and 2:27 marathon and Liz's December 2017 results:

BEPS Target	2:30	2:27	Liz December 2017
AF	44	45	0
PAC	66	68	59
LTCC	118	126	108
ARC3	183	196	0
ARC2	252	269	0
ARC1	329	350	0
ANRC2	N/A	N/A	0
ANRC1	N/A	N/A	0

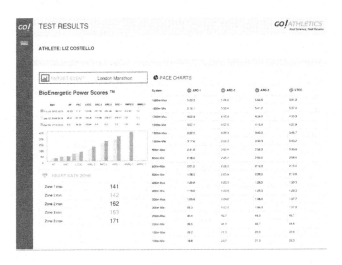

As mentioned earlier, Liz had the PAC and LTCC Systems available, so if I prescribed glycogen depleting training outside of those areas, it would have been unproductive, inefficient, and uncomfortable for her because those Systems were unavailable to adapt to training. The first goals were to use System Priming turn to on her AF, ARC3, ARC2, and ARC1 Systems in order to improve bioenergetic availability and power before building capacity in any of the Systems. Below are an overview and detailed explanation of the recommended training prescription, System, and Phasing, I decided upon to accomplish these physiological objectives for Liz.

Liz's Recommended System and Phasing Prescription:
Phase I- 4 weeks- NAGS (Neuromuscular Adaptation Glycogen Sparing): The primary purpose of NAGS is to set up her daily training load with minimal stress on glycogen stores. In the NAGS phase total volume and intensity in a single session was kept under 75 minutes of Zone 1 to 2 HR. For recovery runs, zone 1 to zone 2 low HR) and for aerobic foundation runs zone 2 low to mid HR with two sessions per week of 10-20 minutes of Economy intervals, 20-60 seconds w equal rest. Her Economy Interval paces progressing each week from her ARC3 down to her ARC1 paces. Her ARC system paces were ridiculously slow, and even though she can run faster for such short intervals, it would have been unproductive for her to do so.

I explained to Liz that the next four weeks would feel uncomfortably slow on her heart rate based runs and workouts.

The uncomfortable feelings were because her Systems were getting reactivated and running in the proper zones to reactivate them would feel inefficient to her, like she was driving a stick shift car for the first time. I needed her bioenergetic power to return to at least six Systems so her body could prepare to build capacity in some of those areas in the later phases. I told her that I fully anticipated that in four weeks, she would recover at least AF through ARC2, and improve her BEPS to the minimal requirements for 2:30 marathon work, while also improving her target system paces by at least 9-15 seconds 1600m per system.

Depending on her actual PPT results we would update her System Phasing and training parameters, HR zones and paces, every 4 weeks. If Liz responded as anticipated, her phasing would be as follows:

Phase II: 4 weeks- ARC3/ARC2: In the ARC phase our purpose was to enhance her upper aerobic system, while increasing her Physiological Range and metabolic adaptability before moving into the final marathon specific (PAC and LTCC) prep work for the final 8-11 weeks. The prescribed ARC3 work was 5 to 10 km in volume. It consisted of 3 to 10-minute intervals with 3-5 minutes rest @ arc3 pace. Her ARC2 comprised work of 4 to 7 km in volume and consisted of 1 to 5-minute intervals with 1 to 3 minutes rest @ arc2 pace. Aerobic (z2 low to mid) runs progressed up to 2 hours during this phase. This phase served as her final ARC power phase to achieve the BEPS requirements for a 2:27 marathon.

Phase III: 7-14 days of NAGS/down week + PPT testing before moving into the final marathon build.

Phase IV: 8-10 weeks: During this time she engaged in PAC (prolonged aerobic capacity)/LTCC (lactate tolerance and clearing capacity) work: The PAC work was performed at Zone 2 mid to high heart rates together with progression runs, zone 2 low to zone 3, of up to 2+ hours. Main System workouts at LTCC (45-75 min @ LTCC pace or zone 3 HR) or PAC (60+ min @ PAC pace or zone 2 mid-high HR).

I was happy to report Liz achieved all the targets on her January 2018 PPT results. We were aiming to reactivate AF through ARC1 Systems, which she did while attaining the BEPS power levels needed for a 2:30 marathon. As stated in the overview of her first PPT, Phase II specifics would depend on her actual test results from the second PPT.

BEPS Target	2:30	2:27	Liz Dec 2017	Liz Jan 2018
AF	44	45	0	40
PAC	66	68	59	65
LTCC	118	126	108	119
ARC3	183	196	0	179
ARC2	252	269	0	256
ARC1	329	350	0	316
ANRC2	N/A	N/A	0	0
ANRC1	N/A	N/A	0	0

Based on Liz's second PPT results, I recommended for Phase II: 4-5 weeks of 4-7 kilometers of ARC2 work, 1-5 min intervals with 1:1 rest up to 3 min max rest @ arc2 pace. At this time, for Liz, I suggested starting at 300m ARC2 intervals and progressing up interval length and total work as long as she could maintain ARC2 pace for each interval for the session. This phase served as her final ARC power phase to gain the BEPS required for a sub-2:30 marathon. During the ARC2 phase, she was to do no more than two sessions of ARC2 work every seven days, preferably spaced out by two or more days. Aerobic Foundation (z2 low to mid) runs could be progressed up to two hours during this phase.

Phase III: 7-14 days of NAGS/down week + PPT testing before moving into final marathon specific capacity build.

Phase IV: 8-10 weeks: PAC (prolonged aerobic capacity)/LTCC (lactate tolerance and clearing capacity): Emphasis on PAC and LTCC capacity work. Aerobic (zone 2 low to mid) and progression runs (zone 2 low to zone 3) can progress to 2+ hours during this phase. Main workouts at LTCC (45-75 min @ LTCC

97

pace or zone 3 HR) or PAC (60+ min @ PAC pace or zone 2 mid-high HR).

After Liz's February 2018 PPT:

Happy to report, significant progress again. Liz was right on target for 2:30 or faster marathon, and now she needed to finish it off with some miles and capacity work in the PAC and LTCC Systems.

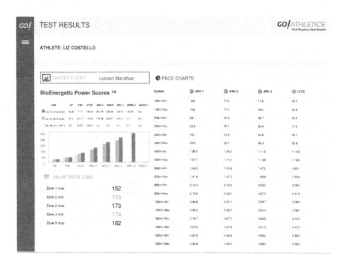

As expected, Liz was a fast adapter, and doing the ARC2 did pull her power up enough to the 2:27 levels and even activated her ANRC2 system, which was pulling her System target paces up significantly. Once she started doing PAC/LTCC work, her bioenergetic spectrum shifted back down to the aerobic end, and her ANRC2 system turned off, intentionally, to build the necessary capacity in the PAC/LTCC Systems. I adjusted System pace charts to account for an ANRC2 System deactivation to complete her marathon prep phase.

Phase IV: 8-10 weeks: PAC (prolonged aerobic capacity)/LTCC (lactate tolerance and clearing capacity): The emphasis in this phase was on PAC and LTCC capacity work. It consisted of aerobic (zone 2 low to mid) and progression runs (z2 low to z3) which progressed to 2+ hours during this phase. Main workouts at LTCC (45-75 min @ LTCC pace or zone 3 HR) or PAC (60+ min @ PAC pace or zone 2 mid-high HR). Adjusted LTCC pace was 5:10.8-5:14.8 min/mile) and PAC pace was (5:35-5:40 min/mile).

I explained to Liz that her bioenergetic power potential was now much higher than a sub-2:30 marathon. In fact it was closer to a 2:20 marathon. Still, it would take more than one marathon capacity phase to execute at those power levels. So, the approach was to go into this marathon prep phase with the goal of a sound platform of capacity work to run sub-2:30 and build on this for a fall marathon. She needed to run controlled efforts, building the miles at the power levels prescribed. For Liz, some of this work would feel unchallenging, nevertheless, she was to focus on more volume and the marathon specific workouts over the next ten weeks. Her marathon specific workouts were prescriber to be between 5:35-6:00 min/mile pace with a top limit of 5:10 with 1-2 sessions of Economy Intervals (1-2 miles of 100-200m strides) after a run at ARC2 pace every seven to ten days.

99

Chapter 7. Engineering Human Performance

Thermodynamics

It is necessary to review the basic principles of thermodynamics to gain a better understanding of the application of bioenergetics in human performance. Thermodynamics is the study of heat and energy, and how it relates to the matter in our universe. The word "thermodynamics" comes from Greek roots meaning heat (Thermo) and energy, or power (dynamics). Energy is the ability to perform work, and when work is performed, heat is produced. Thermodynamics is used by biochemists to understand processes and chemical reactions that occur in living organisms, such as humans.

There are three laws of thermodynamics; the first two are of most interest to biochemists and of great importance as they relate to applied Exercise Biochemistry. The third law has to do with properties of matter at a temperature of absolute zero (about minus 273 °C). It does not apply to biochemistry and is irrelevant to discuss in understanding concepts of applied biochemistry and human bioenergetics.

The First Law of Thermodynamics

The First Law of Thermodynamics states that energy can neither be created nor destroyed. It is commonly known or referred to as the Conservation of Energy law. Conservation of Energy means that the total amount of energy in the universe always remains constant or conserved. However, energy can transform from one state to another. For example, when an engine burns fuel, it converts the energy stored in the fuel's chemical bonds into useful mechanical work and heat. The first law explains key concepts of internal energy, heat, and isolated system work. It is used extensively in the engineering of heat engines.

Different types of fuel can produce different amounts of energy. Still, in any given volume of fuel, such as a gallon or liter, there is a quantifiable and set amount of energy it can produce that will remain unchanged given a set amount of volume. The amount of energy contained in a volume of fuel is based on the chemical properties of its composition, and once the fuel is used up, the engine can no longer produce mechanical work.

Another applied concept of the First Law of Thermodynamics in mechanical engineering states that when all of the fuel's energy is released in a closed system, it remains present. Given this, the total quantity of energy stays the same and must be accounted for in some way. In the case of engines, energy either becomes thermal energy (heat) or mechanical energy (work). For every unit of fuel energy burned in an engine, the same unit of converted energy has to end up somewhere. However, living organisms, such as humans, are open systems. Therefore, the exchange of material and energy within the surroundings is never quite in equilibrium, like that of closed systems. The open human system makes measuring energy availability, capabilities, and efficiency more complex in humans.

The Second Law of Thermodynamics

The Second Law of Thermodynamics is the law of increasing entropy, or level of disorder, randomness, or chaos, of a system. It states that the entropy of the universe increases with every physical process or change that occurs. Still, entropy increases don't have to take place within the reacting system itself.

A bioenergetic balance requires an energy exchange with the outside environment. Biological systems will always tend towards a state of least energy. A vital performance implication of the Second Law of Thermodynamics in human biochemistry and physiology is that naturally occurring processes will always proceed towards the state that has the least potential energy.

The Second Law of Thermodynamics affirms, that the higher the randomness of a system, the higher its entropy. The more organized a system, the lower its entropy.

The Grady Human Performance Theory

Engineers design, build, and create engines that operate with high-efficiency rates for minimal wasted energy and best performance outputs. Designing and building training that creates the best human performance outputs requires the same approach. The most effective human performance outputs also require rapid transfer rates and high efficiency of available energy.

The Grady Human Performance theory grew out of research and analysis on my part that was at first observational, then applied using a distinctive set of measurable variables. Once formulated, I was able to apply it with overwhelming responses and considerable performance improvements across many events and distances.

The Grady Human Performance theory is akin to Carnot's Efficiency Theorem, which is based on the Second Law of Thermodynamics. Carnot's theorem states that if the maximum temperature of a mechanical system, like an engine can be raised, the difference of the input and output temperatures will be greater, thus yielding the maximal possible efficiency of that engine. In general, energy efficiency is the ratio between output energy and input energy. Efficiency refers to how well an energy conversion or transfer process takes place with a minimal loss of energy. However, there is a limit to how efficient mechanical energy can be. The goal is to convert as much of the fuel's energy as possible to work, resulting in more work, and less wasted energy. Therefore, the efficiency or energy conversion of the engine will increase. Likening Carnot's Efficiency Theorem to the function of the "human engine", I theorized that increasing BEPS would raise the conversion rate of the human energy source, (ATP), which, in turn, would increase an athlete's output and thus their performance. Just as there are limiters in engine efficiency (max temperature and the difference in low and max temps), I postulated that the main limiters in human performance would be similar to that of machines and Carnot's Theorem.

The Grady Human Performance theory is my conjecture that human performance has two main limiters that affect total work output and, thus, performance. The two performance limiters are 1. The rate at which one can produce a work output across the bioenergetic spectrum or BioEnergetic Power Scores (BEPS) and 2. Physiological Range (PR). In this case, Physiological Range was defined as the difference between maximum net lactate production and minimum net lactate production. In this case, both BEPS and PR would become vital determinants of any athlete's performance capabilities. In short, the greater BEPS and PR, the greater efficiency and overall work output that can be achieved.

Optimal Physiological Range is accomplished by athletes that have, what I refer to as metabolic adaptability, or ability to utilize substrates across the entire metabolic spectrum. Improving metabolic adaptability will be discussed further in Chapter 11.

To summarize, as with Carnot's theorem, the Grady Human Performance theory states that the optimal way to increase the overall human-system performance output is to increase both the BEPS and PR an athlete can achieve through biochemical and physiological stimulus or training.

Carnot's theorem only applies to engines where fuel such as gasoline is burned. Thus, the fuel burned by humans the increased BEPS component of the Grady Human Performance theory had to be identified. In humans, that fuel was carbohydrates, fats and proteins and their energy-providing derivatives. If ample fuel, which is quantified by the BEPS levels, was unavailable, physiological adaptation and increases in efficiency would not take place. This allowed, The Laws of Thermodynamics, specifically Carnot's Efficiency Theorem, to become applicable in calculating mathematical algorithms for human performance capabilities.

BioEnergetic Power Scores (BEPS) take the Laws of Thermodynamics into account and indicate whether a particular performance outcome is physiologically possible for that individual at that time.

Human Systems Engineering

When I first started teaching human physiology as a graduate student, I have always equated the human body to the most well-designed machine. Our bodies work like machines, with numerous feedback systems, interworking parts, and systems; they can be trained or designed to do almost anything. Something as simple as breathing and sleeping, to running a marathon, biking across France, lifting cars with one's teeth, or walking a tightrope across the Grand Canyon. But how do we know the human LIMITS? How do we know if we are efficient enough or strong enough? How do we know when we have worked hard enough, long enough, or smart enough to ensure optimal performance? Is MORE always better? Is FASTER always better? The question is, for humans, how do we know how much input or training do we need to create the best output for human performance?

There are many different types of inputs, outputs, and efficiencies measured for machines. Still, there have been no direct ways to measure and quantify the efficiency of the most complicated, intriguing machine on earth, the human body! How do we know how much training or input is required to achieve optimal performance or output?

After years of Physiological Profile Testing and human performance analysis, I have found that the most important predictors of human performance follow similar principles to thermodynamic systems, such as machines and engines, and the Laws of Thermodynamics. Every single thermodynamic system exists in a particular state. When a system goes through a series of different states, where a transfer of heat and work occur in and out of the system, and then finally returned to its initial state, a thermodynamic cycle is said to have happened.

In the process of going through training, humans go through a series of different physiological states battling to maintain homeostasis from the increases in heart rate, increase in breathing rates, changes in fuel utilization, and shifts from aerobic to anaerobic metabolism. If given the appropriate stimulus or training, the physiological adaptations will be achieved. A thermodynamic system has a distinctive set of measurable variables and a set of values necessary to uniquely define a system, which is called the thermodynamic state of a system. Similar to thermodynamic systems, the components of producing optimal human performance results include a myriad of bioenergetic and physiological variables. Controlling and organizing the biochemical stimulus applied to the human system can be done via practical, repeatable, measurable, and systematic methods that have an enormous positive impact on all biochemical and physiological responses.

"I trained with Shannon's System Based Training program both as a collegiate NCAA D1 long-distance runner and as a post-collegiate runner. In college, Shannon's system helped me have my best season of competitive running, earning all-Region honors and competing at the national championship race. The

training prescribed by SBT never felt like I was red-lining dangerously and never left me feeling weak from exertion. I was able to train smartly, avoid intensity-related injuries, and save my max efforts for races. As a post-collegiate. I used Shannon's SBT protocol to safely move up in distance and compete in a trail half marathon, which I won. The training never felt overwhelming; it felt accessible, enjoyable, and efficient. I was able to train fully while still working a demanding job in the tech industry, and I showed up at the starting line, feeling incredibly strong and prepared. I've had experience training under several different programs over a 10-year racing career, and SBT produced the best results by far."

-Sydney Harris - Runner

Bioenergetics

What is Bioenergetics and why does it matter? Bioenergetics is a field in biochemistry that involves energy flow through living systems, which includes human cellular and metabolic processes. Bioenergetics notably includes ATP, Adenosine Triphosphate, or the energy source for human movement. Bioenergetics encompasses ATP production and usage along with energy relationships such as the energy exchanged and available to perform activities or functions such as running, swimming, cycling, or rowing.

Applying the First Law of Thermodynamics to human training requires identifying the source of energy. The same principle applies to humans' glycogen stores and glycogenesis, the formation of glycogen from sugar.

In the case of humans, the sources of energy for muscular contraction are chemical. The primary source of chemical energy in the human body is glucose. Its storage form is glycogen, which is a chain of glucose molecules. Glucose is the primary fuel human bodies use to regenerate Adenosine triphosphate or ATP. ATP is the only chemical in the body that can be used to fuel muscular contraction. However, the ATP supply is limited. There is only enough ATP stored to support muscle contraction for four to six seconds. After that, muscle ATP must be regenerated or restored for a muscular contraction to continue.

Glucose is a chemical used to regenerate ATP through a process known as aerobic respiration. Via aerobic respiration glucose and glycogen molecules can produce 36 ATP and 38 ATP, respectively. The number of ATP in a glucose and glycogen molecule is the set amount of energy they are capable of generating via this pathway. This amount of energy is fixed and will remain unchanged per molecule. Humans primarily store glycogen in the liver and muscles. Depending on one's size and daily intake of carbohydrates, the liver and muscles can synthesize and store about 400-700 grams of glycogen.

During exercise, depending on the duration and intensity, glycogen stores can be depleted. The replenishment process of these depleted stores, glycogenesis, takes about 36-48 hours in humans, so if preservation and consumption of carbohydrates are insufficient, glycogen stores will never fully replenish between training sessions. Humans fail to possess a magical glycogen jet pack that will provide an endless energy supply of fuel, in order to insure an adequate carbohydrate availability and preservation for training. Otherwise the athlete will be forced to training partially or fully depleted bioenergetic state. I have found there is a direct correlation with diminished bioenergetic availability and performance potential in most athletes who train regularly. In short, if there is limited bioenergetic availability, the work output, or performance, an individual is capable of generating will always be reduced. No glycogen, no go. This fact is non-negotiable.

Preserving Internal Order

Living organisms, including human biochemical and physiological processes, obey the Second Law of Thermodynamics and operate within it. The human system strives to preserve its internal order by taking free energy from the environment in the form of nutrients, preferably carbohydrates, returning to the environment an equal amount of energy primarily as heat. The more organized the "system" of human biochemical reactions is, the lower its entropy is as well as the predictability of the

110

outcome of its many complex biochemical reactions will be. The components of the human system are extraordinarily intricate, but its responses can be measured, controlled, and organized with high confidence. When evaluating athletes who were in bioenergetic balance but displayed inconsistent performance results, this was due to an arbitrary training stimulus.

Randomized training yields random responses and unpredictable performance results.

If a biochemical stimulus, or training, is disorganized, the higher entropy of the human system and the greater the randomness or unpredictability of the response, or in an athlete's case, the performance outcome. I suggest considering this fundamental concept when programming training as proper fueling or energy intake and recovery are of utmost importance to keeping a balanced bioenergetic state.

From a coaching standpoint, the application of training stressors is the most significant CONTROLLABLE factor of the equation for performance outcomes. So, controlling the controllable is KEY! An acute chaotic state such as academic stress, lack of sleep, or lack of adequate fuel can affect the efficacy of training. Athletes will gain more from training when their homeostatic feedback systems are firing, and their biochemical state is balanced, more organized, and less chaotic. The athlete will be more capable of maintaining homeostasis and still performing consistently under acute adverse "stressful" conditions. Yes, many factors can cause declines in performances, but randomized training is one of the most significant causes of unpredictable performance results in a bio-energetically balanced athlete.

If the athlete is in bioenergetic deficit due to long term outside stressors such as academic stress, lack of sleep, or fuel, coaches can quickly address this by controlling the controllable, which is the training stress. Adjust the training stress to what the athlete is currently capable of handling. Adjusting the training stress to appropriate levels can be easily accomplished by using physiological profile testing data. A coach can quantify an athlete's actual ability to handle training stress using objective, science-based data. If a coach knows the adequate and appropriate training load capacity for each athlete, performance improvements will be expeditious. Following is a letter I received from Luke McCallum, an elite runner detailing his experiences with SBT training.

"My mentality towards running in the past was quite simple. Bluntly, it was "more is more."

Throwback to two years ago. I was juggling full-time work, full-time study, and training at the intensity and load that I thought was expected for someone of my experience level to achieve the goals I had set for myself. It came as no surprise to everyone, except me, that I was always injured, struggled with sleep, and generally didn't feel all that healthy.

This ultimately led me to make a coaching change, which subsequently meant I began utilizing System Based Training (SBT) principles in my training. It was fascinating to see how the data and biofeedback generated from PPT testing would help shape the training that I was to perform. While initially, it was a big blow to my ego when I read my results and recommendations in my first SBT physiological report, (I had to reduce my overall training load and go back to basics for a period), I decided I could look at it one of two ways...

"That's not that good...why should I even continue running..." or I could use the information to understand why I was always getting hurt and make a positive change for the future, and ultimately focus on the room for improvement I had and work towards being a healthier athlete.

Thankfully, I decided to take the second approach.

My coach explained the results and recommendations contained in the report and how we were going to use that to create a plan going forward. He also told me how we would utilize repeated PPT testing to determine how I was tracking towards my goals and adjust the plan as needed.

Since that first test, I have been tested another five times, with the most recent test producing by far the most exciting results. I have gone from having five functioning energy systems with relatively low capacity to now having 7 (out of 8 Systems) with significantly increased capacities. But what does that mean to *me? I'm healthier, happier, and my performance has dramatically improved as a direct result of developing the application of variables in my training program. And all this has been possible while doing LESS than I did in the past when I was always injured, and my performance had stagnated.*

The training that has led to this improvement has been varied and enjoyable, while still being focused on getting me healthy and functional.

I've also noticed that by having a training program that facilitates restoring health and functionality in conjunction with performance, rather than just pure performance outcomes, it has given each day a purpose, which is something I have lacked in the past. Now I can enjoy each run/training session/rest day because I understand how it is building towards maximizing both my health and performance collaboratively.

It is easy to say that I am a big fan of SBT as within the past year, my half marathon personal best has improved from 1 hour 17 min, down to 1 hour 10 minutes and 25 seconds, and in general, I feel the best I have for ages!

I can confidently say my "more is more" days are firmly behind me, and I'm excited to continue my journey towards maximizing my full potential using PPT testing and the SBT training principles."

-Luke McCallum- Runner and Neuromuscular & Massage Therapist Sports Lab New Zealand

Chapter 8. Winning With Science-System Based Training

The power of big data is that mathematically, we are closer to a more precise answer. That power is why System Based Training is so effective. Data on over 100,000 athlete samples gives us enormous insight and certainty as to the outcome athletes can expect from a Physiology First Training Model.

Physiology First Training Model

The ultimate quest for most coaches and sports scientists is to determine which training plan design will create the optimal physiological responses as well as performance outcomes across various sports and events. Due to the variability of each athlete's physiology and substandard training plan design, adaptation to training can be unpredictable and occur at different rates. For example, longitudinal data from 2010-2014 indicates that NCAA 5,000 meter runners improve performance time annually by a mean rate of only 0.7%, with 65% of these runners improving by less than 2% or less in any one year. There is no agreed-upon optimal training plan design by sports science experts and coaches in regards to intensity, volume, phasing, workout type that will work for all athletes. Therefore, the idea of a truly individualized training plan design in team settings seems like an impossible challenge. Some methods and models have emerged that tout individualized training by using metrics such as "lactate threshold," VO2 max, performance marks, or event focus. The periodization of such models remains static, failing to prescribe training based on an individuals' actual physiological status. Such approaches are where the individualization ends and generalization begins. In these popular physiological assessments, training plan design based on a single data point, and then standardized formulas are applied to develop training paces, zones, or efforts. Using these standardized methods assumes the physiological capacities and bioenergetic outputs of each athlete are the same at each percentage of those values. These assumptions are short-sighted and disregard an individual's current bioenergetic status, potential, and capacities.

Some suggestions have been made to prescribe training based on muscle fiber type shifting. The reality of measuring exact percentages of muscle fiber type, fiber type distribution, contribution, recruitment, and ability to shift is unknown and impractical. Some suggest that the use of maximum lactate values can indicate an athlete's muscle fiber type breakdown, but in reality, genetics is static, and physiology is dynamic. One's maximum lactate values can vary by up to 160% from one test to the next. Therefore, knowing if the current sample is indeed an indication of genetic potential or an athlete that is in Bioenergetic deficit remains unknown unless there is comparative PPT data of an athlete in an untrained state. Designing training with intentions to shift fiber type is a waste of time but also unreliable, immeasurable, and an unpredictable approach to develop individual athletes for performance purposes. Muscle biopsies can only determine local muscle fiber type distribution and not representative of fiber shifts in all muscles. A muscle biopsy is a surgical procedure where typically a 25g needle is inserted into a muscle, usually the thigh, and a small amount of tissue is removed from the subject. While most report the pain as being minor, this is an extremely invasive procedure. A muscle biopsy is an undesirable method to measure an attempt to measure fiber type shift information. Assumptions are that a maximum lactate reading is an indication of muscle fiber type, but this is also an extremely invalid assumption and method. Maximum lactate readings are unreliable unless numerous data points are measured during variations in training load and stimulus.

Overload of training, inappropriate training stimulus, or inadequate recovery could be the cause of any performance stagnation or declines. Still, when it comes to eminent competition, there is no time for guessing games. While time trials or field testing provides only one data point, these

118

methods are unable to answer the underlying question of "why," nor can these methods offer a feasible training solution to the coach or athlete. Fortunately, more sports scientists, coaches, and athletes are open to the idea of using lactate testing. Many are realizing that objective biometric data provides insight versus hindsight approach. Insight data answers questions as to why performances are becoming stagnate or which training plan will work best for physiological development. Coaches or athletes with minimal experience using blood lactate as a tool often assume that decreases in lactate readings at the same velocities are always a positive response to training or indicative of improved fitness. This assumption and application of the use of lactate measurements often lead to performance declines and bioenergetic deficit.

Among the sports science community, it is agreed that individualization in training periodization will create optimal outcomes for performance. Yet, true individualization of training prescription is rarely carried out by coaches or sports scientists. Measurement of individual physiological information and applying that data correctly is the only accurate way to individualize and differentiate training responses. If physiological information is unknown about the respective athlete, the coach is making a best guess effort, which is often an impairment to development and performances. A meta-analysis on this topic points to the overwhelming consensus that the training stimulus designed to optimize an athlete's current physiology is the most feasible and effective means for positive performance outcomes.

SBT testing and training methods evaluate aerobic and anaerobic energy contribution changes along the human bioenergetic spectrum. After analysis of thousands of athlete samples, SBT testing protocols have been developed for many sports and events to evaluate the necessary bioenergetic variables to meet performance goals. Analysis of Physiological Profile testing data is via proprietary algorithms, which yields the individual's BEPS, heart rate zones, power zones, and training paces. The basic premise of System Based Training (SBT) is that objective biometric data is used to determine individualized training parameters that will achieve the desired physiological effect and objectives with great certainty. The key variables that are used in the SBT physiology first model to achieve those objectives include net lactate, velocity, power, heart rate, capacity, work-to-rest ratio, and trainability. SBT primarily serves as a tool for coaches to quickly implement complex scientific data into daily training programs/plans.

The four-step process of SBT implementation is seam-less and straightforward in the athlete's regular training regimen. The first step is biomarker analysis via physiological profile or anaerobic rate capacity testing. Testing is performed entirely in the field using standardized sports-specific protocols and methods. Each protocol evaluates the complete energy spectrum required for peak performance. The second step is data processing, analysis, and translation into actionable parameters. The third step is to advise coaches and athletes on how to assimilate data into an existing daily, weekly, and yearly training program. The fourth step is on-going support and consulting for practical data application and implementation. The SBT physiology first model enables coaches to manage training load, reduce the incidence of overload related injuries, and maximize performance results.

System Based Training is a physiology first, positive, response dynamic training model.

The SBT physiology first model uses only known variables that are objectively measured and quantified. These variables are then applied using various scientific principles of biochemistry, physiology, exercise physiology, metabolism, and nutrition to achieve specific training and performance objectives. The development of the SBT methodology was via testing and analysis of biochemical and physiological data from heterogeneous athlete testing samples across a myriad of sports. Many coaches ask if SBT is useful for any sport. The answer is yes. Human physiology is the same, no matter what sport or event one engages in. The SBT physiology first model can address one's physiology and performance in individual sports such as triathlon, running, swimming, rowing, or team sports such as soccer, basketball, and football.

The strength of response to each System does have a genetic component based on the individual's physiological makeup of the muscle fiber type of each athlete. The SBT physiology first model works to optimize the physiological power and capacity of each athlete. SBT makes no guarantees that each athlete will be a World Class athlete, but it will enable each athlete to achieve their maximum potential.

The System Based Training physiology first model is highly effective in producing performance improvements but also in providing predictable physiological and biochemical responses. In addition to using the using scientific principles of biochemistry, physiology, exercise physiology, metabolism, and nutrition, the SBT model is based on the Laws of Thermodynamics.

The Second Law of Thermodynamics is the law of increasing entropy or level of disorder, randomness, or chaos of a system. SBT methodology decreases the level of disorder of biochemical stimuli applied to the body by taking a focused, systematic, single stimulus block training approach. . Block training design uses training blocks and sequencing in four to six-week segments where the training stimulus is highly concentrated and focused on a minimal number of physiological, biochemical, and neuromuscular stimuli. On the other hand, most training philosophies or designs use a mixed system training approach. A mixed system approach is where coaches apply many different training stimuli in hopes that some respond with improved performances. Jack Daniels' is famous for touting the "eggs against a wall" approach, in which the goal is to develop many physiological areas simultaneously in hopes that one of those stimuli creates a response. Few individuals respond positively to this approach, and, to the contrary, most respond negatively. The fact of the matter is, the rates of improvement among athletes, on the whole, are weak. Talented athletes, those that are classified as, physiologically and genetically gifted, have highly adaptive physiological and biochemical responses to a training stimulus. These rarities will shine through almost any program design, as they have atypical reactions to any stimulus. Therefore, training design and observational research based on the top athletes' performances are limited in their ability to yield similar results in most athletes. In reality, the majority of athletes have inferior response rates and outcomes when compared to genetically gifted athletes. Block training suggests limited stimuli and a more rational sequencing of each training block.

8-12 weeks

The early stages of SBT development involved tracking athlete physiological profile results and training response rates every eight to twelve weeks. The results displayed that a focused, systematic, single stimulus block training model produced a 98% predictability rate, stronger physiological response rate, and significant performance improvements across multiple running events.

Over 12 weeks from initial Physiological Profile testing and implementing System Based Training prescriptions, the following performance results were achieved across ten different Division I track programs:

-1 minute 40 seconds to 2 minute 20 seconds improvements (5-7%) over 10,000 meters

-1 minute 20 seconds to 2 minute 10 second improvements (3-5%) over 5,000 meters

-22-44 seconds (3-6%) over 3000 meters

-5.4 to 8.6 second (3-6%) improvements over 1500 meters

-4.68-15.49 seconds (3-10%) over 800m

In short, NCAA runners who adhered to SBT training prescriptions had a significantly greater performance improvement rate in 12 weeks over the typical NCAA runner improvement rate of less than 1% in a year. Joe Campagni, Director & Head Coach of the Monmouth University Men's and Women's Track & Field and Cross Country Teams was kind enough to state the following about his experience with the SBT approach to training.

"Shannon has a unique and thoroughly scientific approach to help coaches and athletes understand their current physiological profile. Not only does she administer the lactate testing and provide great data for each person, but she also provides precise and customized recommendations to help guide the training for each person. Her efforts with our team the last few years have allowed us as coaches to eliminate much of the guesswork of workout planning and have elevated the performance of our athletes in cross country and track."

-Joe Compagni- Director & Head Coach, Monmouth University Men's & Women's Track & Field and Cross Country

Dynamic Periodization

It is necessary that training and periodization of training be dynamic and based on individual physiological data to maximize the performance of each athlete.

In sports, the job of the coach or performance specialist is not only to plan the training of athletes and teams to match event or sport demands but also to have the athletes peak to perform their best at the right time. Planning of training is known to sports scientists and coaches as periodization. The concept of periodization was derived from Hans Selye's model known as the General Adaptation Syndrome or GAS. Selye was an early pioneer, circa the 1940s, in studying biological stress responses and his GAS model describes physiological responses to stress into the following stages: 1.an Alarm stage, 2. A Resistance stage, and 3. An Exhaustion stage. Proper periodization intends to provide the body with good stress, which is also referred to as eustress, that will yield positive adaptations and physiological responses. By adhering to cyclic stress and rest phases without exhausting the body, the incidence of distress or adverse physiological reactions will be minimized.

In general, periodization models follow periods of high load, volume, and intensity of training, followed by periods of low load or rest. These are called microcycles. In the 1950s, physiologist Leo Matveyev and sports scientist Tudor Bompa further expanded and applied standard periodization models with great success for Soviet athletes in the 1960 Olympics which then spread to the Eastern Bloc and Romania. In the 1970s, more individualized periodization models using physiological data started came into use. There are typically three cycles in periodization models: the microcycle (about seven days), the mesocycle (two weeks to several months), and the macrocycle (overall period such as a year or two.) Indeed, it has become common to plan four-year macrocycles for aspiring Olympic athletes and teams.

It is widely accepted by sports scientists and coaches that training should be organized and planned for the season or year, but there is little agreement on which model of periodization is superior. There are many opposing stances of every sport as to which training method or periodization is most effective. At the same time, many studies and reviews question the validity of traditional periodization models, which are primarily static, due to their multitude of assumptions about human response and adaptation, especially with no physiological data to support these assumptions.

John Kiely (2012), published an extensive literature review on periodization, in which he states that the limitations of most periodization models are due to the real-world scenarios concerning the variability of human responses to stress as well as the dynamic nature of physiology. Kiely believes most generic periodization models are insufficient for addressing complex biological systems. For this reason, Kiely suggests that

126

periodization models need to evolve to align with elite practice and modern scientific concepts. He also suggests that design and training should be responsive, and variation integrated into periodization. Kiely, John. (2012). Periodization Paradigms in the 21st Century: Evidence-Led or Tradition-Driven? International journal of sports physiology and performance. 7. 242-50. 10.1123/ijspp.7.3.242.

Most periodization models based on numerous assumptions and scenarios may work in theory and controlled settings. However, In the real world of athlete performance, there are no controlled settings. All athletes competing in the real world are uncontrolled subjects, and coaches are unable to manage and control all aspects and variables that contribute to their ability to respond to training stress. Since 2006, the System Based Training model has done just as Kiely suggested. System Based Training is an individualized training prescription that addresses real-world scenarios of the variability of human responses to stress as well as the dynamic nature of physiology. The main goal of SBT training prescriptions, System and Phasing, is to ensure that each athlete is bio-energetically balanced, and the training prescription is appropriate for positive physiological responses. The goal of the SBT physiology first model of periodization is to improve an athlete's physiological range, maximize performance, and also to reduce the risk of injuries. Even if the athlete is in a team sport, the team will be a much stronger unit with more bio-energetically balanced individuals contributing to that unit.

When planning periodization, it is crucial to keep in mind that emphasizing one System too frequently or for too long a period, will most often lead to Bioenergetic Deficits and underperformance. Each System is essential to an athlete's overall performance, and each System will play a unique role in that performance. Training to perform best at peak season requires a thorough knowledge of the human body's response to training, and the type of training necessary to perform the specific sport or event. Regardless of the sport, bioenergetic and physiological analysis is essential to understanding the role each System plays in performance, as well as its present status. Bioenergetic, and physiological profile data will provide greater insight on how to effectively periodize training for each athlete.

Many coaches spend copious amounts of time and energy planning daily, weekly, monthly, and even yearly training plans. Some plans come with much detail and thought, but most often, the athletes' adherence to the coach's intended plan or objectives are lost through the day to day perceived exertion efforts of the athlete. Allowing athletes to follow perceived exertion efforts will be only minimally effective in achieving performance gains. In 2018, physiological profile testing of over 2,700 athletes across 15 different sports (80% Division I, 10% high school, and 10% recreational level) displayed almost no correlation between the RPE (Rate of Perceived Exertion) and net lactate values.

Individualizing training has obvious advantages for the individual athlete. It can, however, be equally effective for competitors in team sports. However, despite its advantages, individualizing the training for members of a team can be a time-consuming and challenging task. However, you should know that group training is only minimally effective in achieving optimal physiological adaptations for each individual athlete in an interacting group activity. A single training philosophy, volume level, intensity level, or frequency level applied to all athletes is ineffective in creating expected or predicted performance responses and accounting for physiological differences that can explain training adaptation inequalities in each athlete is often an impossible task when no physiological information is measured.

It is widely accepted among the sports science community that individualizing training and periodization will create optimal outcomes for performance. Yet, true individualization of training prescription is rarely carried out by coaches or sports scientists. Measurement of individual physiological information and applying it correctly is the only accurate way to individualize and differentiate training responses. If physiological information is unknown about the respective athlete, the coach is making a best guess effort, which is often an impairment to development and performances. SBT evaluates Biochemical Systems and applies Training Systems to produce measurable, predictable, and repeatable training responses for each individual and for each bioenergetic system that athlete must use during competition in a particular sport. Each System contributes to the physiological capacity he or she can bring to the task at hand. Therefore, each athlete's biochemical Systems must be measured by various SBT physiological testing protocols. Once testing is complete, the exact individualized training parameters can be set, which I refer to as, System and Phasing. All training parameters and recommendations are repeatable, measurable, and yield PREDICTABLE results. Training parameters can include one or more of the following: heart rate, pace, watts, time, volume, and frequency. Each athlete's training parameters are individually calculated, utilizing testing information to create optimal training adaptations.

The following comments by an elite coach and athlete in the sport of rowing describing the feelings of many athletes who have used SBT training protocols.

"It took a while for me to appreciate the impact that Physiological Profile Testing and System Based Training had on me as an athlete and as a coach. I simply did what my coaches prescribed, and I ensured the execution of training for athletes that I was coaching based on what the head coach had communicated to me and others on the staff. When starting out, in both cases, I didn't really question what I was doing. I met Shannon a few years back while still racing, but it took until I was finally operating as a head coach to be able to implement her methods. This was 35 years after I started rowing and after 25 years as a coach. The insights she was able to provide have paid huge dividends in the performance of my athletes.

I am curious by nature as I still consider myself a student of the sport. Understanding individuals' physiological profiles and addressing physiological deficiencies in the pursuit of being fully functional through the targeted implementation of power-based training made a lot of sense to me. Since applying SBT to several athletes and crews that I have coached it has resulted in several PRs, world championship trials victories, and a few trips to the world championships. I am now using that same methodology on the high school athletes that I coach.

Meanwhile, as an athlete, I was able to leverage emerging technologies, like GPS and power meters, and apply the testing and training on myself. I nailed my power targets on power-based days and nailed my heart rate targets on recovery and aerobic foundation days. After half a dozen PPTs and 18 months of training, I can say that I have seen quantifiable improvements, both in terms of the power and speed, that I have not seen in 20 years.

I no longer have to guess what my athletes need to do or what I should be doing. I no longer have to rely on applying training plans that are nonspecific and too general in content.

When Shannon told me about this book, I can say that I will be first in line for its purchase."

-Christopher McElroy- Elite Rowing Coach & Master's Rower

System and Phasing

Most often, the flaws with many training models or philosophies are that one training aspect of a periodization plan is emphasized and applied to all athletes, season after season. The System Based Training physiology first model is an entirely dynamic periodization model that steers the planning process of training based on the most recent physiological profile testing results. SBT is prescribed primarily in four to six-week training blocks that emphasize training a particular System. The four main dynamic periodization concepts used when prescribing the "System and Phasing" aspect for each athlete are Main System Workout (MSW), Priming System (PS), Bridging System (BS), and Focus System (FS).

Main System Workout (MSW)

The Main System Workouts, or MSW, refer to the primary energy system and workout type that is emphasized in that phase of training. One to two MSW is prescribed every 7-10 days during any phase of training. Each MSW during any phase will be for the same System for each workout to maximize the biochemical dosage and response rates. The MSW can be any of the 11 biochemical and training Systems: R&M, NAGS, AF, PAC, LTCC, ARC3, ARC2, ARC1, ANRC2, ANRC1 or HMCT.

Priming System

System Priming is prescribed to create a biochemical and physiological catalyst before full System phasing or overload stimulus of any one energy system. System Priming is a training method that I developed over approximately eight years of field testing and tracking the results on hundreds of unique athletes of both sexes, all ages, and widely varying ability levels. Implementing System Priming demonstrated a stronger reactionary biochemical and physiological response during subsequent System phases. System Priming is conducted during the off-season, early season, and competitive season training phases. The prescription of the Priming System focus and order is dependent on individual Physiological Profile testing results, goal event or competition, and time until goal event or race.

The Priming System refers to the System used during that phase as the System primer. A Priming System can be the primary training System for a phase but is most often a secondary training System used during some other phase of training or competition period. System Priming serves several purposes, such as 1. Igniting Systems that are currently unavailable 2. "waking up" a System that has been unstimulated or trained for a few months or more that is going to be the Main System focus in the next phase. 3. Off-season training. The Priming System can be any of these seven biochemical Systems; PAC, LTCC, ARC3, ARC2, ARC1, ANRC2, or ANRC1.

Bridging System

The Bridging System or Systems are ones that "bridge" the gap or training blocks between phases leading to the peak competition. Prescribing System and Phasing is unique to each athlete's scenario that includes but is not limited to the following considerations: 1. Recent PPT results, 2. Amount of time until peak competition or event, 3. Type of event or sport in which one competes. Depending on the number of phases prescribed between a PPT test and peak competition, there can be none or multiple Bridging Systems prescribed. For each case or athlete scenario, the System stimuli are optimally placed to accelerate biochemical responses for each athlete and event in which an athlete is training. Bridging Systems can be any of the 11 biochemical and training Systems; R&M, NAGS, AF, PAC, LTCC, ARC3, ARC2, ARC1, ANRC2, ANRC1 or HMCT.

Focus System

Focus System is the System that is ideal and should be strongest for peak performance. The Focus System can be determined but is unlimited to the following considerations: 1. Event or sport in which one competes, 2. Amount of time until peak competition or event, 3. PPT data.

System Based Training aligns with applied biochemistry and physiology principles of stress, adaptation, and response that will optimize one's physiology. Each System is essential, and each System will play a role in performance. Although training athletes to perform at their best during peak season requires a thorough knowledge of the human biochemical and physiological processes and responses to training. SBT provides coaches with a sound and simple platform to yield those results. Understanding the roles and physiological basis of each System will provide greater insight on how to develop proper training for any event.

Below is a case history of a high school cross country team comprised of 22 boys and girls. The team had initial Physiological Profile testing in August 2015. The runners were grouped by bioenergetic availability and recommended System and Phasing that was intended to optimize their 5-km times by the end of October 2015. At the end of September 2015, this team tested for a second time after five weeks of System Based Training. Some of the runners were mid-distance specialists (800m/1 mile), and some were long-distance specialists (3200m/5000m). The training prescription given to this group in August had little change in total training volume during the five weeks between PPT testing sessions.

The chart displays the percentage of training in each category of their total training System emphasis and volume breakdown over five weeks. The maximal percentage of "lactate threshold" or "tempo" training over the five weeks for all the groups was ten percent. Yet, the improvements in their "lactate threshold" pace, calculated from the velocity at four mmol, ranged from 23 to 71 seconds with the LEAST increase coming from the group that performed the highest amount of "lactate threshold" training. Their results are described by the coach in the following testimonial.

# Of Athletes in Group	Group Average Improvement @ 4mmol pace	Zone 1 (R&M)	Zone 2 (Aerobic Foundation)	LTCC HR or Intervals	NAGS Intervals	ARC2 Intervals	ARC1 Intervals
1	27 seconds	32%	27%	10%	17%	0%	14%
12	23 seconds	32%	27%	10%	17%	0%	14%
8	26 seconds	44%	23%	0%	23%	0%	10%
1	71 seconds	42%	23%	0%	35%	0%	0%

"Over the past two XC seasons at Northern Valley Demarest, we have used the services of Shannon Grady and her team. Last year our team ran their fastest 5K team average in school history while sustaining "NO" running injuries. All that was from only three months of Shannon's System Based Training. This season we decided to plan the whole summer and fall with Shannon, and so far, all of the results are both promising and exciting. Some athletes have already progressed further than anticipated. We are projected to "peak" out at our state sectional championships and are on pace for some flying times.

Besides the training aspect of things, you also get a level of professionalism and care that goes way beyond what you pay for. Shannon and her team are there for you, helping you 24/7 as if your athletes are their athletes. You get the feeling that they are on your coaching staff and not an outside consulting group. Being a national & collegiate level athlete, I always knew this type of training existed. Still, I never thought it would have become available for high school athletes both at a price point and convenience. For the price, you get a season's worth of personalized workouts, paces, and advice. For convenience, they come to you for all of the testing sessions. I can't wait to see the results at the end of the season. In the meantime, we have a lot of quality training ahead of us."

-Michael Theuerkauf- Northern Valley Regional HS- Head Coach XC and T&F, 25 years of coaching combined at the NCAA Division 1 and High School levels.

Conestoga High School Senior Andrew Marston had his first PPT in June of 2013. Andrew's followed the recommended off-season System Priming throughout the summer. Andrew had his second PPT test in August of 2013. At this time, many of his teammates were tested and adhered to the SBT parameters their coach' implemented into their daily training. That cross country season, Andrew dropped his 5-km cross country performances from 16:30 in 2012 to 15:50 in 2013. Andrew was tested several more times during his Junior year; December 2013, April 2014, and June 2014. Throughout his Junior year track season, his BEPS scores continued to improve, and his times dropped significantly in all events: 1600m- 4:41 to 4:20, 3-km- 10:30 to 8:45, 3200m- 10:00 to 9:17. Andrew flawlessly followed his SBT training prescription and had a spectacular senior year. During the cross country season, he finished 3rd place at the State Cross Country Championship meet and broke the previous course record.

	AF	PAC	LTCC	ARC-3	ARC-2	ARC-1	ANRC-2	ANRC-1
Apr 01, 2014 01:01	43.17	77.09	145.17	217.32	298.12	383.65	0.0	0.0
Dec 14, 2013 01:01	45.4	73.83	134.42	202.13	276.37	352.24	424.49	0.0
Aug 26, 2013 01:01	39.4	65.18	126.32	189.41	257.2	329.5	0.0	0.0
Jun 06, 2013 01:01	0.0	62.3	120.57	184.55	250.06	314.79	0.0	0.0

Test Result Chart

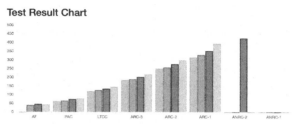

Andrew's BEPS Profiles 2013-2014

The Conestoga High School, cross country team, not only saw success in their top runner but in all of their runners, except for one who was unable to participate in running several days a week. Below is an overview of their team cross country results for three seasons.

Conestoga HS (Pennsylvania AAA) 2012-2014 Cross Country Season Comparison (top 5 runners score, a lower score is ideal)

Conestoga Boys	2012 XC	Before SBT	2013 XC	1st Season w team on SBT	2014 XC	2nd Season w SBT
District I Championships	Place	Time	Place	Time	Place	Time
TEAM PLACE	17th		2nd		1st	
1st Runner	52nd	16:30	18th	15:50	2nd	15:18
2nd Runner	59th	16:36	28th	15:56	14th	16:04
3rd Runner	72nd	16:44	38th	16:11	21st	16:18
4th Runner	163rd	17:17	40th	16:13	41st	16:32
5th Runner	201st	17:28	49th	16:21	74th	16:52

PA STATE XC CHAMPS   2012-NO QUALIFIERS   2013- 7TH PLACE TEAM   2014- 2ND PLACE TEAM

To note, during the 2014 cross country and 2015 track seasons, 12 High School boys who were on SBT went sub-16:00 in their 5-km cross country races and sub-4:20 in the mile during the track season. Eleven of the 12 boys were running 4:30 or slower during the 2014 track season.

The assumption with many training programs is that there are a variety of approaches can make improvements in novice athletes such as high school runners but that such changes are very difficult to achieve during their college years. Below is the case history for one men's Division I cross country program.

Division I Men Team Cross Country : 2012: Prior to SBT

Race	Team Score	Team Place	5th Runner	7th Runner	9th Runner
Conference	114	5th	31st	44th	59th
Regional	57	2nd	19th	50th	n/a

NCAA Championships   475   20th

2013- Began SBT 6/2013

Race	Team Score	Team Place	5th Runner	7th Runner	9th Runner
Conference	36	1st	14th	16th	24th
Regional	34	1st	12th	39th	n/a

NCAA Championships   415   18th

2014- Began SBT 6/2013

Race	Team Score	Team Place	5th Runner	7th Runner	9th Runner
Conference	23	1st	11th	12th	21st
Regional	41	1st	15th	16th	n/a

NCAA Championships   230   7th

Chapter 9. Bioenergetic Deficit

PPT analysis of athletes who exhibit many symptoms of the "Overtraining Syndrome" demonstrate significant declines in their bioenergetic availability. When evaluating PPT with diminished bioenergetic availability, the question always becomes which factor is responsible for the negative physiological response: training, fueling, recovery, or iron levels? Sometimes the answer can be many of these factors. PPT has a diagnostic value in monitoring declines in bioenergetic availability. Regardless of the cause, I refer to these declines as Bioenergetic Deficit. A Bioenergetic Deficit occurs when, over time, more energy is being expended for training and life stress than in being replaced through fuel intake and rest. In time, a bioenergetic deficit will certain metabolic systems to become unavailable for use during training and competition. Systems can shut down from any of the factors mentioned, and the severity of the System shut down depends on how long the negative factor has been present.

Identifying and Overcoming Bioenergetic Deficit

PPT analyses of athletes who experience some level of underperformance, demonstrate some degree of Bioenergetic Deficit and System shutdown among all. I believe a Bioenergetic Deficit exists when an athlete has less than six Systems operating during training. The astounding similarities among all these cases are classified along the continuum of Bioenergetic Deficit as Stage I- Mild, Stage II- Moderate, and Stage III- Severe.

Stage I- Mild Bioenergetic Deficit - only 4 to 5 Systems Available
Stage II- Moderate Bioenergetic Deficit - only 2 to 3 Systems Available
Stage III- Severe Bioenergetic Deficit - only 1 System Available

Recovering from each stage of Bioenergetic Deficit will require a minimum of seven to ten days of decreased training stress. Depending on the level of severity and how adherent the athlete is to adjustments in training load as well as addressing other factors that can contribute to Bioenergetic Deficit, recovery can take up to a year or more when an athlete reaches Stage III.

The standard training prescription I have developed to address various stages of Bioenergetic Deficit is called a Non-Glycogen Depleting Training Protocol (NGDT) in which athletes avoid training so long or intensely that they deplete their muscles of glycogen without replacing the same from day to day.

Athletes typically have immediate performance improvements within 10-14 days of NGDT training without changing any other factor that could contribute to Bioenergetic Deficit. Training volume and intensity is the primary factor that a coach or athlete can control in the uncontrollable real-world setting of athletics. Adjusting training volume and intensity to address various stages of Bioenergetic Deficit is simple. The basic guidelines for training prescription based on PPT data is seven to ten days of NGDT for each unavailable System.

NGDT generally consists of the following;

1. Two full rest days every seven days. ,75
2. The total training volume should be less than 60 minutes per day.
3. Training intensity should be effortless and conversational or Zone 1 to Zone 2 low heart rate.
4. Incorporate 5-20 minutes of Economy Intervals 1-2 times per week. Economy Intervals are 20-60 second intervals @ a prescribed System intensity with 20-60 seconds rest.

Following is a description of the experience of Amy Horst, Head Coach for Cross Country and Track at Loyola University with overcoming bioenergetic deficits in her athletes using these guidelines.

"Shannon's research has removed the "guesstimating" from my coaching. I initially began working with Shannon when I had a high performing athlete that was stagnant in racing. We were quickly able to identify her deficiencies, make the appropriate adjustments to her training, and see improved race results in the same season. Since then, I've been able to use the testing results to keep my athletes healthy, fast, and continuously improving!"

- Amy Horst - Loyola University Head Coach- Cross Country & Track

Bioenergetic Deficit Signs and Symptoms

DING DING!

The most common signs of Bioenergetic deficit during training and competition are reductions in acceleration, sustained power or velocity, and a reduced endurance capacity, as well as inability to handle the physical load of training as evidenced by plateaus or declines. Other signs are both psychological and physical in nature; lack of confidence, reduced recovery, and increased injury.

Experiencing Bioenergetic Deficit symptoms can be scary, confusing, and frustrating. Consequently, deciding upon the path to take from a training standpoint can be very complex; Do more? Do less? Rest more? Rest less? Knowing which will be the answer to turning performances around can be quite difficult to determine.

There are no truly defined parameters or conclusive laboratory findings to diagnose "Overtraining Syndrome." that can be found in the scientific literature at the present time. Often, athletes are deemed "over-trained" if they have had several months of plateaued or declined performances. Also, they have the following symptoms: general feelings of fatigue, restlessness, elevated resting heart rate, abnormal stress hormone levels such as cortisol, and lack of motivation to train. These common symptoms are some of the many possible reactions from chronic bioenergetic imbalance and the inability for an athlete to return to homeostasis, a stable condition of bodily functions which enables optimal human performance, upon cessation of exercise. What is even more frightening is that the common symptoms of "overtraining" appear after an athlete has been in a bio-energetically imbalanced state for at least two months. On the other hand, physiological profile testing can detect an acute bioenergy imbalance in as little as 7-10 days.

The topic of "Overtraining Syndrome" has always been an area in which diagnosis is unclear and treatment is misguided. Accordingly, the detrimental effects on athletes can be acute, benign, or long-term with serious health implications. I do feel there is a case to be made for blaming inappropriate training rather than "overtraining" as the cause. "Overtraining" implies that the adverse effects are simply due to the sheer volume and intensity of training when the volume and intensity might have been appropriate for that athlete if they were in a state of bioenergetic balance. It is next to impossible to put limits on a volume and intensity that can apply to all athletes at all times. Monitoring an athlete's bioenergetic status can ultimately yield a physiology first, positive response training model. In other words, it is often the case that the training stress that resulted in the appearance of overtraining would have appropriate if the athlete had been bio-energetically balanced. For example, if an athlete paid attention to replacing the necessary fuel, iron, ferritin, hematocrit, hemoglobin, red blood cell count, or other life stressors through the use of recovery periods, the same training load might have been appropriate.

If the training load is inappropriately prescribed for an athlete's current bioenergetic status for two months or more, they will display common signs and symptoms of "overtraining." While the actual amount of training might be appropriate for an athlete, who was adequately fueled and rested. There are no real indicators and symptoms other than PPT testing to inform a coach or athlete, until it is too late that their homeostasis has been disrupted. Repeatedly putting out more energy than one is putting in will cause the body to be bioenergetically limited in its ability to restore glycogen (carbohydrates stored in the muscles and liver) daily.

If humans want to go fast, they need glycogen; there is no magic or trick to escape that metabolic fact. Glycogen is the high octane of performance racing. Glycogen is the primary fuel that drives fast performances, and without proper recovery (up to 36 hours between intensive or prolonged efforts), glycogen stores are unable to be replenished fully. If an athlete performs intense training uses glycogen as its chief source of fuel, more than two to three times in seven days, their glycogen "tank" will chronically depleted.

Fortunately for humans, unlike our cars, we can continue to work if we run out of high-octane glycogen, this is because, the human metabolic system will adapt for SURVIVAL purposes. A human can survive for a while without adequate glycogen stores by using other, less rapid, sources of energy for fuel, such as fats and proteins.

High performing fast humans are genetically superior in that they have two dominant energy systems that are strong, the anaerobic, as well as the aerobic system. Genetically inferior humans only have one strong energy system, and that is the aerobic system. Fortunately for our species, everyone has a metabolism that can adapt to inferior slower burning, longer lasting aerobic fuel so we can survive if glycogen runs out. Still, our performances will surely and noticeably suffer. Humans have no choice but to move slower if there is no high-octane fuel, or glycogen, available. No amount of mental willpower can overcome this deficit and make one move faster. Unfortunately, competitive humans have the mental willpower to ignore and override less subtle physiological signals to SLOW DOWN until there is a bigger, more noticeable problem such as a string of bad performances, or an injury that eventually forces us to stop.

An athlete being bio-energetically imbalanced for as little as seven days can cause measurable physiological, metabolic, and performance changes. Depending on the severity and extent of the energy imbalance, it may take months of zero activity and proper fueling to return to normal physiological and metabolic status. The first sign of energy imbalance is typically repeated slower performances. Unfortunately, competitive human nature is to think we need to work harder when this happens. Mistakenly believing we should do more to get faster. But, in actuality, we need to do less to get back in bioenergetic balance in order to run faster. The typical reaction when athletes are underperforming leads to more work and digging them deeper into Bioenergetic Deficit.

In such cases, the training load is the primary variable that coaches and athletes CAN control. It is, the most challenging stressor for human homeostatic feedback systems. If the training load is beyond what an individual can handle given their current metabolic state, the ability for the body to maintain homeostasis post-workout will become increasingly more challenging. If one is experiencing "overtraining" symptoms, he or she is likely in Bioenergetic Deficit and needs to restore glycogen and adjust current training volume and/or intensity. It may also be a good idea to record a seven-day food log to ensure that enough calories and carbohydrates are being consumed to fuel training demands.

Below is a case history of a high school-aged club swim team. Upon initial evaluation, ALL 28 athletes had moderate to severe levels of Bioenergetic Deficit. The chart shown below indicates the number of athletes and their total number of available Systems during both their April 2018 and August 2018 PPT.

Between April and August 2018, ALL of the swimmers in this group improved their performances over their respective events. Below are the results for the various groups of swimmers tested as well as the Systems evaluated, together with the Phasing, and daily training prescription recommendations followed by this team as well as a statement from coach Ian Gross detailing their experience.

Total Functioning Systems By Athlete

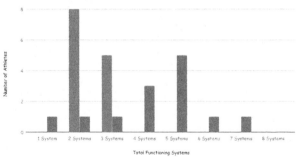

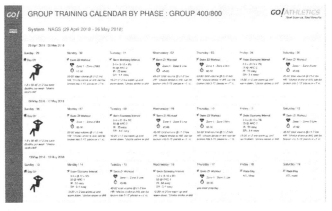

"Working with Shannon and her team was very easy. Shannon explained her test protocol and the results we gained from them. Shannon is very easy to communicate with and always answers my questions about what an athlete's weaknesses or strengths are. Our results this summer were outstanding, the swimmers all improved, and they got a better understanding of why we worked on specific parts of their training the way we did. They are already asking when we are going to do the testing again. As a coach, learning what type of training is needed by each athlete is a must for their development. The implementation was flawless and very flexible to utilize within *our training program. Having a sports scientist help guide you with each athlete is a big help and a great experience."*

-Ian Goss, Head Coach Stingrays Swimming

Chapter 10. Diagnostic Performance Monitoring

Think It's Mental? Fortunately, Blood Don't Lie!

Are human performance limits in the mind? Are performances all mental? Fortunately, blood doesn't lie. Numerous cases of underperformance or regression of performance indicate otherwise; bioenergetic availability and power are the primary determinants of performance capabilities. Most often, the athletes possess sufficient mental fortitude to perform, but it is a lack of bioenergetic availability and power that limits performance capabilities.

Case in point, meet "Billy," a rising, young athlete with an abundance of talent. As a freshman, Billy had run 4:15 in the mile and was a determined, coachable, and focused athlete. Billy's parents always tried to provide or find the best resources for their children, so they engaged a professional coach to help their son get to the next level. Billy, like most motivated kids, did everything his coach prescribed in training, never missed a workout, and was encouraged to push himself as much as he could every training session. Much of his training regimen included "tempo" runs and workouts geared towards improving his "lactate threshold." Over the next 18 months, his performances stagnated and ultimately declined. He was working harder than ever yet failing to see the fruits of his labor pan out in races.

During his junior year indoor track season, Billy was struggling to run under 4:30 in the mile and under 2:00 in the 800m, which were slower times than he had achieved during his freshman year. His times were atypical of the progression of improvement that would be expected of a boy as he physically matures throughout his teenage years. His parents, now concerned with reasons why their son's performances may be deteriorating, sought out professional mental health services as well as doctor's advice for several months with no change in Billy's performance status. Although to the average runner, these times may sound good, for Billy they represented declining performances which both he and his coach attributed to mental weakness. Billy was blamed for being physically and mentally weak because the coach believed he was not working hard enough.

In March of 2014, via a family friend, Billy's parents reached out to me for physiological profile testing. Billy's initial PPT test results had him operating at 50 percent of the necessary bioenergetic power needed to run his previous times. Billy was in Mild Bioenergetic Deficit, with four Systems currently available (see corresponding BEPS chart). Unfortunately, his upper end aerobic or anaerobic Systems, ARC or ANRC Systems, which were required for optimal performance in his events, were unavailable at this time. Billy's physiological evidence confirmed that no matter how hard he tried or how tough he was, his body was unable to run faster in training and competitions. He was missing the necessary bioenergetic availability of Systems that his body needed to perform the same mile times he had been capable of in the past.

Billy's 2014 BEPS charts

Test Result Chart

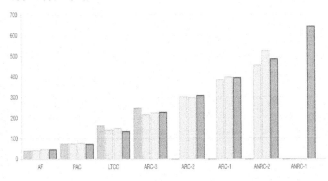

	AF	PAC	LTCC	ARC-3	ARC-2	ARC-1	ANRC-2	ANRC-1
Dec 13, 2014 01:01	47.02	73.22	135.54	227.78	308.76	395.42	487.26	644.99
Sep 10, 2014 01:01	48.18	78.7	150.26	225.92	299.33	398.78	526.76	0.0
Jul 09, 2014 01:01	43.76	76.07	142.54	216.7	302.5	386.28	456.52	0.0
Mar 08, 2014 01:01	40.57	74.95	163.68	248.17	0.0	0.0	0.0	0.0

Billy's PPT results showed that he was training in a glycogen depleted state too frequently, which, in turn, led to his inability to access certain necessary parts of the bioenergetic spectrum. After analyzing his March 2014 PPT test, I prescribed Non-Glycogen Depleting Training (NGDT For Billy to perform faster, he needed at least six Systems available before he could resume full System focused training. His training prescription and the results he obtained were as follows:

Initial PPT March 2014: Mild Bioenergetic Deficit, 4 Systems available
Phase I: NGDT
Physiological Objective: Achieve bioenergetic balance with at least six Systems available
Duration: 4 months- March-June 2014
Parameters: <60 min total volume per day, Economy Intervals @ ARC-1 paces
Performance Changes: March-June: 1-mile: 4:25 → 4:14

On July 9, 2014, I conducted Billy's second PPT. As anticipated, Billy had bioenergetic balance, no Bioenergetic Deficit, seven Systems available. In addition, he had doubled his physiological range and increased his BEPS, 300-450%, in three Systems that were necessary for optimal middle-distance performances. His second PPT test revealed the following changes which resulted in a new training prescription for Phase II:

Second PPT July 2014: Bioenergetic balance, No Bioenergetic Deficit, 7 Systems available
Phase II: AF + System Priming
Physiological Objective: Increase AF capacity, prime LTCC and ARC Systems in preparation for the cross-country season, increase bioenergetic availability
Duration: 8 weeks
Parameters: 2 weeks: AF + EI

Two weeks: AF + ARC-1

Two weeks: AF + LTCC

Two weeks: AF + ARC-2

Performance Changes: NONE in his major events due to summer cross country training

On September 10, 2014, I conducted Billy's third PPT. As anticipated, Billy had bioenergetic balance, no Bioenergetic Deficit, seven Systems available, maintained his physiological range and increased his BEPS in all seven Systems. This resulted in a move to Phases III, IV, and V which included 5 weeks of ARC-1 training in the Phase III, followed by 5 weeks of LTCC training in Phase IV and 5 more weeks of ARC-2 training Phase V. His results for the cross-country season are listed following phase V.

Third PPT September 2014: Bioenergetic balance, No Bioenergetic Deficit, 7 Systems available

Phase III: ARC-1

Duration: 5 weeks

Parameters: 4-6 km; 200-800m intervals @ ARC-1 pace

Phase IV: LTCC

Duration: 5 weeks

Parameters: 5-8 km; 600m-1-mile intervals @ LTCC pace

Phase V: ARC-2

Duration: 5 weeks

Parameters: 5-8 km; 400-1200m intervals @ ARC-2 pace

Performance Changes: Cross Country 5k: 16:30 → 15:30

On December 13, 2014, I conducted Billy's fourth PPT. As anticipated, Billy had bioenergetic balance and no Bioenergetic Deficits, with all eight Systems available. Additionally, he had increased his physiological range and increased or maintained his BEPS scores in six of his eight Systems. These results are detailed below.

Fourth PPT December 2014 PPT Bioenergetic balance, No Bioenergetic Deficit, 7 Systems available

Performance Changes: His first Indoor Mile of the previous season took place in January 2014 at which time he ran a 4:32-mile. Later, in December 2014 he ran a considerably improved time of 4:13 in the mile. Billy also ran an Indoor 1000m in 2:28 in December 2014.

Billy had four PPT evaluations between March 2014 and December 2014. As evidenced in his BEPS charts, his bioenergetic availability increased markedly through the use of an individualized System Based Training (SBT) plan. The recommended training prescription yielded measurable, reliable, and predictable results across the necessary Systems for enhanced performance for his various events.

Billy's continued performance improvements directly correlated to his Bioenergetic Power Scores and physiological range increasing and improved his performances throughout March

2014 and June 2015 when he achieved a top ranking for high school boys in the United States. Subsequently, Billy had a stellar Senior year with incredible performances. He ran a 15:30- 5-km Cross Country race, which was a one-minute improvement from the 16:30 he had run in his Junior year. He also ran 9:10 for 3200m, 2:27 for 1000m, and a remarkable double of a 4:09 - mile and 1:51- 800m at the State Indoor Championship meet. His 4:09.56 was a new State and Meet Record.

Interestingly enough, the typical "lactate threshold" testing analysis would have determined that Billy's "lactate threshold" decreased 30% over the nine months. Common recommendations based on significant decreases in "lactate threshold" most often suggest more work at "LT" and assume negative performances correlated to a decline in "lactate threshold" velocity. Yet, in-depth System analysis reveals that his BEPS in five other Systems increased 200-610 %, and predictably, his performances improved.

These results indicate that BEPS scores may be more predictive of improved performance than the common practice of lactate threshold testing followed by training designed to improve only this one measure of the total metabolic spectrum. BEPS scores represent an extremely valuable and insightful metric for objectively quantifying physiological progress, regression, or stagnation as well as determining the area of physiology that will yield a positive response for that individual at that point in time.

An athlete's current bioenergetic availability and power can be measured and quantified through BEPS test scores. BEPS data can actualize which biochemical stimulus or training parameters are appropriate for a particular athlete to improve bioenergetic availability and power in any area of the bioenergetic spectrum. Thus, it appears that creating specific responses to increasing bioenergetic availability or power is a matter of providing the appropriate biochemical stimulus for that person at that time. By measuring and applying individual bioenergetic data to daily training, there is greater certainty of yielding intended and predictable physiological and performance outcomes. Without bioenergetic data, there is no way of knowing whether an athlete is currently able to adapt to the training stimulus given. There are no non-invasive tests, devices, apps, or algorithms that can measure or know whether a training stimulus is appropriate, productive, or unproductive for an athlete at any time. Providing a known stimulus that an athlete is capable of adapting to, no matter what that stimulus is, will create a positive and highly predictable biochemical response. In short, if the appropriate biochemical stimulus or training is applied, then intended response biochemical and physiological adaptation will occur.

Following is another case study for a Division I male 800-meter runner and the results he achieved with SBT between August 2014 and October 2014. The PPT and accompanying BEPS charts display that the intended physiological responses were achieved in the form of increased bioenergetic power and availability. by specifically addressing each bioenergetic System via the appropriate training stimulus.

Pre-Season PPT Analysis:

August 14, 2014/Mid-Season PPT Analysis: October 17, 2014

	AF	PAC	LTCC	ARC-3	ARC-2	ARC-1	ANRC-2	ANRC-1
Oct 17, 2014 09:09	43.04	72.13	141.61	216.37	302.01	398.27	0.0	0.0
Aug 14, 2014 09:09	0.0	46.7	92.75	148.01	213.13	289.99	0.0	0.0

Test Result Chart

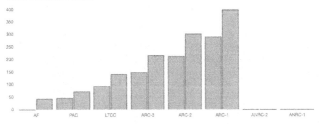

Goals: Increase AF + ARC System Priming

Duration: 6 weeks

Training Prescription Parameters:

2 weeks: 4-6 km: Economy Intervals (EI), @ ARC-2 pace

2 weeks: 4-6 km: Economy Intervals (EI), @ ARC-1 pace

2 weeks: 4-6 km: ARC-2 Intervals, 300-400m @ ARC-2 pace

Following is a copy of a statement I received describing Dylan's experience.

"After my race today, I want to give a big thanks to all the coaches I've had the honor of training under and all the words of wisdom I've been given. Especially to my coach, Chris Tarello, and Shannon Grady, for all the hard work these two people have

poured into my training to get me to where I needed to be. I wouldn't be the athlete I am today without their help. Thanks, Guys. #gosystembasedtraining"

- Dylan Capwell, after taking 2nd in 1:46.7 at the 2015 NCAA Indoor Track Championships

Basing training programs on BEPS data allows for the development and identification of areas of physiology where most significant adaptations are for each athlete at the current time.

BEPS can and will change throughout different training phases, and understanding the relevance of each System's power score for the event is fundamental to improved performances. The comparative analysis of BEPS is a valuable tool for understanding current performances and an individual's physiological adaptation. Implementing BEPS data into subsequent training includes what bioenergetic outputs one can handle at any given time and will reduce the risk of bioenergetic deficit and injury.

There have been no real scientific conclusions or overwhelming agreement among sports coaches as to which training parameters create the most optimal performance outputs in all humans or how to measure these variables repeatedly, accurately, and effectively. Training can be quite a game of trial and error for coaches and athletes, in which the only assessment tool is a race outcome. BEPS testing has a distinctive set of measurable variables and can assess all the variables of the human bioenergetic systems that impact human performance outputs and rate limiters. BEPS enables a coach or athlete to make objective, science-based decisions when developing training plans.

"I started working with Shannon back in 2013. She opened my eyes to a type of training that is based on science and fact, not guesswork, a type of training that is specific to each athlete, and a type of training that is dynamic. The information she provided me was absolutely instrumental in producing our 2016 YMCA State Championship team as well as several other nationally recognized performances over the 2015 - 2016 season. Upon moving to Georgia, I asked Shannon to come down and work with us at the Columbus Aquatic Club. Again, the information she was able to provide our team with was second to none. We were able to calculate individual heart-rate zones for aerobic and recovery training, as well as training paces for specific energy systems. The information gives the coaches a glimpse into the physiology and potential of each athlete. It provides a training protocol that allows us to continue to train within a team setting, which is important within the constraints of this sport. With her help, we were able to produce several Georgia State Champions and Futures Champions this summer and secure a handful of Junior National and Senior National cuts."

- Andrew Beggs - Columbus Aquatics Club

Return To Play

It is impossible to outperform one's current biomechanical and physiological limits.

PPT data implemented for training optimization and preventative measures for performance declines is unparalleled amongst human physiological performance evaluations. PPT data provides benefits due to its ability to objectively evaluate the etiology of performance decreases or plateaus as well as provide a more precise means to determine appropriate training prescription after injury or a "return to play" plan. When athletes are repeatedly asked to perform tasks outside of their current biomechanical or physiological limits, the incidence of injuries increases. Currently, no standardized evaluation protocols are being used that can provide objective measures that quantify and prescribe the training volume and intensity that is appropriate for athletes' actual biomechanical and physiological limits.

Return to play after injury or illness should include a specific and individualized plan that prescribes the appropriate training volume and intensity for that athlete to resume training and return to 100 percent participation safely. Most often, this process is left up to the athlete or a coach to determine how much work the athlete can handle on any given day. PPT data is beneficial for sports medicine staff and coaches to prescribe appropriate training volume and intensity for athletes who are medically cleared to return to play. Following is a testimonial received from a working mother who is also a TeamUSA triathlete.

"I started working with Shannon in 2013 when I was recovering from an injury. She not only helped me to manage my recovery with massage but also helped me to continue training despite it. Since then, she has guided me into being a much stronger, more balanced, and more competitive triathlete. I've continued to improve, and each year I set the bar higher, knowing that with Shannon's training and guidance, I will meet my goals. As a mom to two young kids and a business owner, I don't have all day to train. Shannon gets that. So my training plans reflect the fact that I have a limited amount of time and help me to make the most of it. Working with Shannon has been a super experience, and I'm looking forward to continuing to learn, grow, and improve as a triathlete under her guidance!"

-Rosemarie Miller, Self-Employed, Two-time TeamUSA triathlon member

To provide a case study of an athlete who was able to make a successful return to competition after illness, let's take the case of High School Senior Brandon Hontz. I had been coaching Brandon Hontz since his freshman year, with his main emphasis being the sport of triathlon. Brandon competed for the high school cross country, swim and track teams, and also in triathlons during the summer. Brandon made steady progress throughout his high school running, swimming, and triathlon career, and heading into his senior campaign he was fresh off a 19th place finish at the ITU Sprint World Championships, a 15:30 at the cross-country state meet, and an indoor mile personal best of 4:23.

Something felt "off" during his state meet, and he struggled to finish the race. I told Brandon to take two weeks off, and then I performed a physiological profile test in March 2017. His PPT resulted in Stage I- Mild Bioenergetic Deficit with only four Systems functioning. After reviewing Brandon's historical PPT data, I insisted he get a mono test, which turned out to be positive. Due to Brandon's deficit of two Systems, I prescribed NGDT training for the next four weeks with the intent of igniting his ARC2 and ARC1 Systems again so he could potentially run fast in the outdoor season. I re-tested Brandon in April 2017, and as expected, he regained his two ARC Systems, and I resumed his usual training volume and intensity. This took place six weeks before his outdoor state meet. Brandon ended up with a personal best performance of 4:14 in the mile at the end of May 2017.

	AF	PAC	LTCC	ARC-3	ARC-2	ARC-1	ANRC-2	ANRC-1
Apr 18, 2017 02:02	38.13	64.97	127.23	194.93	284.82	397.99	0.0	0.0
Mar 17, 2017 10:10	44.82	76.56	149.16	238.18	0.0	0.0	0.0	0.0
Nov 20, 2016 10:10	44.15	71.72	137.42	216.58	310.13	0.0	0.0	0.0
Jun 03, 2016 03:03	47.41	78.65	148.34	228.05	311.01	0.0	0.0	0.0
Mar 08, 2016 03:03	36.82	61.83	125.79	199.11	288.79	0.0	0.0	0.0
Jun 04, 2015 01:01	33.9	57.2	115.24	179.57	252.62	0.0	0.0	0.0
Mar 09, 2015 01:01	32.89	56.78	107.82	162.23	226.83	0.0	0.0	0.0
Sep 15, 2014 01:01	37.21	64.61	117.97	177.25	247.32	317.01	0.0	0.0

Test Result Chart

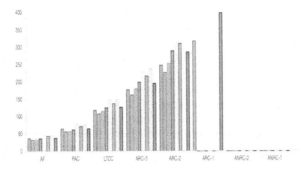

Female Athlete Triad, Is It Just For The Ladies?

The Female Athlete Triad consists of three main symptoms, including low energy availability, menstrual dysfunction, and decreased bone mineral density. Susceptibility to the Female Athlete Triad is the most common condition that differentiates male and female athletes mainly since the loss of a menstrual cycle (amenorrhea) is the major red flag for this syndrome. Loss of a menstrual cycle is impossible in males since they have no menstrual cycle. Still, the other symptoms and subclinical disorders of hormone dysfunction, fatigue, low bone density, etc. that wreak havoc on many female athletes, especially endurance athletes, can also affect many male athletes.

Many of the conditions caused by hormone dysfunction can be long-term conditions that are irreversible. Much of the research on the Female Athlete Triad shows that one's "energy" availability is the source of this syndrome, and exercise is blamed as the culprit. Unfortunately, no research measures or quantifies actual "energy" availability besides looking at dietary or caloric intake versus exercise output.

Chronic energy imbalance, Bioenergetic Deficit, will ultimately lead to Female/Male Athlete Triad and many other conditions associated with inappropriate training stress.

Strong, healthy athletes need to address all of the factors that contribute to the conditions associated with low energy availability and Bioenergetic Deficit.

More athletes need to be monitored for acute and chronic Bioenergetic Deficit to avoid many of these potentially harmful conditions. Currently, the treatment for athletes who show signs of low dietary energy availability is to increase their caloric intake. This approach, although necessary, only addresses one side of the energy equation. Achieving bioenergetic balance requires the energy out from training, and stress must be less than or match the energy in through proper fueling, and rest. If only the energy in is addressed, the underlying cause, energy out, will still be present, and the condition will only continue.

The following is the case of an Elite Female 1500-5000-meter runner, Mary. During Mary's 2012 track season, she was experiencing training plateaus and significant decreases in her performances over three months. Her 5000-meter race performances declined from 15:30 to 16:05 and then again to 16:45 between February and April of 2012. Mary was receiving training guidance from her college coach, but for the most part, she was designing her training based merely on how her body felt. After her season of sub-optimal performances, she decided to take a seven-week "break" to recover. Mary's subsequent training was three to four weeks in what she described as "easy," after which she said she felt "recovered" and ready to resume training for the fall racing season.

Mary's initial PPT analysis on September 12, 2012, resulted in Stage II- Moderate Bioenergetic Deficit. Even though Mary felt as if she was ready to resume her typical training volume and intensity, her physiology showed otherwise. For Phase I, Mary was prescribed four to six weeks of the Non-Glycogen Depleting Training Protocol (NGDT) to increase her total for available Systems to at least five to six in that she could begin full System training for the upcoming season. Mary's NGDT consisted of less than 60 minutes of zone 1-2 moderate runs and 10 to 20 minutes of Economy Intervals (EI) @ ARC-2 pace. Phase II was prescribed as four to six weeks of 2000 to 5000-meters of LTCC intervals for her MSW (Main System Workout) sessions.

Mary returned to competition on November 17, 2012, and ran a personal best time over an 8000-meter road race of 26:29. Mary continued to improve her bioenergetic availability to seven Systems as evidenced by the results of her second PPT on December 18, 2012.

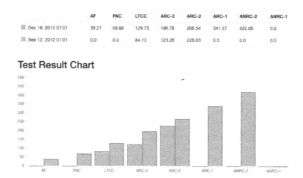

	AF	PAC	LTCC	ARC-3	ARC-2	ARC-1	ANRC-2	ANRC-1
Dec 18, 2012 01:01	39.27	69.98	129.73	196.78	266.54	341.27	422.05	0.0
Sep 12, 2012 01:01	0.0	0.0	84.13	123.28	228.63	0.0	0.0	0.0

Test Result Chart

Mary's MSW training prescription following her second PPT was 3000 to 6000-meters of ARC2 intervals. Mary competed in two races throughout this phase, and both resulted in personal bests: February 2, 2013 she ran an 8000-meter cross country race at 27:45 and February 9, 2013, she completed a 3000-meter indoor track race in 9:03.

Chapter 11. Improving Performance Potential

"Eating alone will not keep a man well; he must also take exercise" Hippocrates.

Hippocrates evangelized the concept of a whole-body approach to health that advocated both healthy eating and exercise. Since then, science has demonstrated ad nauseum the correlation between positive health effects, physical activity and nutrition.

Improving Metabolic Adaptability and Bioenergetic Availability

Improving bioenergetic availability must take a whole-body approach in order to integrate the principles of exercise biochemistry and the effects of nutrition on metabolism and performance. Total fuel and macronutrient intake are essential to ensure one has adequate bioenergetic availability. Sufficient fuel, macronutrients, and calories are necessary to what I refer to as, metabolic adaptability, or ability to utilize substrates from the entire metabolic spectrum, to adapt to training stress.

To keep a check on fueling and bioenergetic availability status, I recommend the following;

1. Keep food logs for at least seven days to track total calories and carbohydrate intake.

2. Never train with a depleted glycogen supply.

3. Depending on one's body weight, take in at least 100-200 calories of solid or liquid food within 1 hour before every workout.

4. ALWAYS refuel. Eat or drink at least 100-200 calories with at least 50 grams of CARBOHYDRATES within 1-hour post-workout.

5. Aim for 7-8 hours of sleep each night. Add bonus energy with frequent 10-20 minute power naps.

6. Get a Physiological Profile Test, which will give exact measures of current bioenergetic status.

Improving metabolic adaptability and bioenergetic availability are both dependent on adequate substrate availability and application of the appropriate biochemical stimulus or training stress. Applying the proper training stress is the one variable over which coaches have direct control. Training parameters such as volume, intensity, and work to rest ratio, should be adjusted according the System or Systems that are being targeted. Each of the 11 training Systems has unique parameters which can be applied to achieve desired bioenergetic and physiological effects with great certainty. The following is an account of a World Class competitive roller skater and her experience overcoming Bioenergetic Deficit and challenges with injuries and training.

"As long as I can remember, all I've ever wanted to do was to compete at the Artistic Roller Skating World Championships.

After years of hard work in the sport I loved, my dream came true when I was 17 years old. In 2011, I represented New Zealand in Brasilia, Brazil. The competition was the experience we all hope for. I skated a personal best and felt proud of myself. That's all I could ever ask for as an athlete.

The following year, I once again competed at the World Championships in my hometown, Auckland. Another highlight of my skating career. It felt so surreal to come out with 7th place of 15 skaters — a massive improvement from the previous year. However, I was now in my first year of university and struggling with the workload. I was taking out my frustration during training, and this resulted in training way too hard and not eating enough. I was desperately trying to keep up with the slim aesthetic of our performance sport. Looking back now, I can say I was very fit, but it was not a sustainable lifestyle. I was driving myself into the ground without even realizing it. That year, I failed one of my first-year papers. I was on the pill, but even so, my periods were never regular. I was also in pain all the time. I took it as a sign that I was "working hard." Still, later, I came to find out I had tendinopathy in my dorsiflexor muscles (tibialis anterior and extensor digitorum longus). This injury was not diagnosed until it was too late. I was afraid that I would be forced to take time off to heal, which would interfere with my goals. I was constantly in pain every time I skated, but I kept training through it. These were all very alarming warning signs that I failed to notice because I was too "focused" on training as hard as I could.

In 2013, I was training harder than ever and still studying full time at university despite the pain and under-eating. I was not giving my body a chance to recover from my training. I went to Taipei to once again represent my country and compete at the World Championships. This time, the competition fell right in the middle of the exam period, which meant I had to sit an exam while I was overseas. During an official practice at the World Championships, two days before I was due to compete, I sprained my ankle. I was devastated. I had been working all year for this moment only for it to be taken away by an ankle sprain. I went to the official competition physio, and she questioned my pain and swelling because it didn't add up to just an ankle sprain. I thought my swollen continuously tendons were only part of being an athlete. That's when she told me I had to see a physio when I got home. But of course, I still finished the competition.

Reactions like this are, unfortunately, all to common among elite athletes. Following in another example of the dangers of training with bioenergetic deficits.

When I got home from Taipei, my feet and ankles were swollen continuously. I wasn't even able to walk without pain. I was left frustrated and angry that my body had let me down. I had no choice but to stop all exercise for a few months, and I quit skating for at least six months. For someone with difficulty around food, being forced to slow down was hard. I went through a phase of feeling lost as I had to face who I was without skating. I had never stopped skating for that long before.

Over the next few months, I suffered six ankle sprains just from walking down the street. The pain in my feet and leg muscles was not improving. I tried physios, acupuncture, sports physicians, and nothing was improving. In the end, I found deep tissue massages and rest to be my saviors. I got a massage every 2-3 weeks for a year, and I stopped competing seriously. I learned to find the fun again in my sport once I took the pressure away. I slowly fixed my eating habits and learned to eat properly. I finally allowed my body to recover from the previous years of abuse. In March 2015, I came off the pill after nearly five years of taking it. I was 21 at the time and had never had a regular cycle even when I was on the pill. I didn't get my period until August that year, and then after that, it took another three months to come again. I now know how detrimental that would have been to my overall health and recovery.

Late in 2017, I decided I was ready to compete internationally again, but I swore I was not going back to my old destructive mindset. That's when I asked my friend Mathew Mildenhall to step in and help me train for the 2019 World Championships, but this time using Shannon's Physiological Profile test and System Based Training.

I have been using Shannon's way of training since 2017. At first, it felt like I was not doing enough. It felt like I would never be able to get fit enough to perform a 3-minute routine by doing less. It seemed counter-intuitive. But that's precisely what happened. I cut down on how much I was training and started training smart instead. I learned to diversify my training intensities so that not every day is 100% intensity. This has allowed me to be on skates more than ever before! I'm happy to report that I'm still off the pill and have very regular periods even now in the middle of my hard training cycle. This way of training has also taught me that it's ok to take time off when life gets hectic. And that it's in your best interest to take time off as that's better than pushing through the training program and crashing later on.

Previously, thinking about competing at worlds again left me with a feeling of dread, and the training overwhelmed me. But since working with Matt and Shannon, I'm feeling ready to kick ass more often than not. Following a plan excites me and reassures me in a way I never thought it would. Because the program is tailored to me, I know I have to follow what is on there. It takes away a lot of the guesswork I previously had to do and gives structure to my training. It also keeps me in check when I feel like I "want to do one more rep" because I know doing more now may mean I don't recover enough to do my next training."

-Macarena Carrascosa (Mac)- New Zealand Artistic Roller Skater

Fallacy of High Fat/Low Carbohydrate and Caloric Restriction Diets

A common trend of late that is touted as the "secret to success" for enhancing endurance is training and eating to become "metabolically efficient." What exactly does this mean? The term efficiency combined with metabolism sounds like an athlete's dream, especially to an endurance athlete. After all, efficiency means to "save energy without waste or unnecessary effort." Well, this admittedly sounds like it could be the key to success for endurance performance; go as long as possible and waste as little energy as possible. Unfortunately, however, the training and fueling methods recommended to achieve metabolic efficiency come at a cost, a high cost, which happens to be poor performances.

The basic premise of "Metabolic Efficiency" is that athletes can "teach" their bodies to burn more "fat" for a longer period of time by eliminating or severely reducing carbohydrate intake to a bare minimum. Theoretically, this will cause the body to adapt and convert "fats" longer instead of converting carbohydrates to produce energy for exercise. The theory sounds promising, but the reality of physiological adaptation is that there are limits to human efficiency and rate limiters to performance output which prevent it from happening. As discussed, ad nauseam, human metabolic efficiency, and metabolic adaptability follow the principles of the Laws of Thermodynamics. These performance limiters can be explained via the Grady Human Performance theory that the two performance limiters are BEPS and Physiological Range. Although increases in metabolic adaptability will yield an increase in performance, It is impossible to achieve metabolic adaptability, optimal BEPS scores, and Physiological Range for ANY sport or event without adequate carbohydrates and total calories.

This component of the Grady Human Performance theory, as with Carnot's theorem, ONLY applies to engines where fuel is burned.   On the other hand, The Grady Human Performance theory ONLY applies to athletes in which ample fuel or substrates are available. If sufficient fuel, namely carbohydrates and substrates from carbohydrates, are unavailable, physiological adaptation and increases in efficiency will diminish. Overloading one system by training it too much or only providing one metabolic substrate will create a state of reduced overall system efficiency, Bioenergetic Deficit, and performance output.   Human substrate utilization and physiological adaptation in each System are limited and calculable.

Here is a prime example of inadequate macronutrient fueling. James, a recreational long-distance triathlete took an online DNA test in January of 2017 that reported his preferred fuel for exercise to be fat. Therefore, he was advised to consume a high fat/low carbohydrate diet.   NEWSFLASH, no DNA tests are needed to realize ALL HUMANS PREFERRED FUEL SOURCE IS FATS! Human metabolism has evolved for survival, and fats provide more energy for survival. Consequently, it is innate in our DNA to prefer fats for SURVIVAL purposes. After adjusting his diet to the below macronutrient breakdown, James continued his usual long-distance triathlon training, after adjusting his diet to consume high amounts of fat and fewer carbohydrates. James' performances deteriorated over the next 18 months as well as his state of well being. James always felt tired and lethargic.

James reached out to me for advice in August of 2018. His PPT in August of 2018 resulted in Moderate Bioenergetic Deficit, with only two Systems available. James was overloading his low end aerobic (AF and PAC) Systems by only providing one primary substrate, fat, and this decreased his mid to upper end aerobic (LTCC, ARC3, ARC2, ARC1) System efficiency and ultimately its availability. The loss of each System caused an overall decrease in metabolic adaptability, bioenergetic availability, efficiency, performances, and eventually reduced his ability to maintain homeostasis at rest, even though he still had his "fat-burning" System available.

Given James' level of Bioenergetic Deficit, I also performed a nutritional and expansive biomarker evaluation, which revealed inadequate macronutrients and substrates, mainly carbohydrates, required to fuel his training needs as well as disturbances to his hormone production capabilities. As referenced in Chapter 2, lactate is a crucial hormone signaling molecule. So if one's ability to produce lactate diminishes, then hormone production will also decline.

I prescribed James eight weeks of NGDT before he could resume his normal training load. James had already committed to an Ironman event eight weeks after his PPT and wanted to do it "for fun" so I told him he could do the race, but he could only do NGDT leading to the event. James completed the Ironman feeling strong and improved his performance from the previous year, while he was doing what most would consider inadequate training for an iron-distance event.

James' Macronutrient Profile

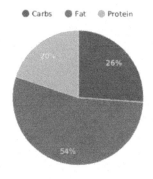

James' BEPS Chart

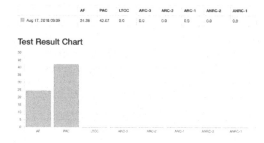

No matter how much of one type of fuel a person consumes, the human metabolic system is unable to infinitely adapt to that fuel without a cost. In human metabolic adaptation, if there are limited substrates available for consumption, then there will be conservation somewhere. Limiting the availability of one or more substrates will slow down overall metabolic efficiency and a reduction in metabolic adaptability, which results in the conservation of energy and the decline of performance capabilities.

Basic science shows the human body as the most well-designed machine there is. The human body has innate feedback systems to alter metabolism when survival is necessary, or fuel is short. The survival metabolic adaptation that occurs when the body burns "fat," free fatty acids and ketone bodies because inadequate carbohydrate stores are available is known as ketosis, a metabolic state where energy supply comes from ketone bodies instead of carbohydrates or glucose. This metabolic adaptation to "fats" or ketone bodies will inhibit glycolysis because a molecule of fat needs metabolites from glycogen or glucose molecules to be present in order for it to undergo glycolysis. If you remember, glycolysis is the name for the metabolic process that converts glucose or glycogen molecules into pyruvate and hydrogen ions while yielding ATP. As discussed previously, a metabolic pathway used in glycolysis that is highly productive for yielding ATP is the ETC (electron transport chain), where oxidative phosphorylation occurs. One glucose molecule yields 32 ATP molecules in oxidative phosphorylation, which accounts for about 90% of the total ATP synthesis in humans.

In ketosis, without carbohydrates such as glycogen, stored carbohydrates, or glucose, in the blood, aerobic and anaerobic metabolism is unable to take place sufficiently, and a person's ATP output will be reduced. ATP, as stated earlier is what creates human power or bioenergetic outputs. Bioenergetics notably includes ATP production and usage, along with energy relationships, such as the energy that is exchanged and available to do work. Humans are living systems that produce ATP primarily from organic sources, namely carbohydrates, via oxidative phosphorylation. To the contrary, free fatty acids yield far fewer molecules of ATP because many tissues are unable to convert these fatty acids to synthesize ATP. Thus, the metabolism of glycogen molecules will have a higher bioenergetic value than the metabolism of ketone bodies. So, the rate at which a human can produce both work (i.e. running speeds) and the capacity to do work (the length of time one can run at a given speed) will be drastically reduced, when "fats" are metabolized for energy instead of carbohydrates. Metabolizing "fats" is also a much slower process than is metabolizing energy from carbohydrates. Metabolism of "fats" will yield The slow metabolic rate for combustion of fats will also be accompanied by slow running speeds and diminished performance outputs.

The bioenergetic cost of relying on fats contrary to carbohydrates is such that the velocity of work rates will be reduced. When the human body has inadequate carbohydrate stores, it must resort to inferior bioenergetic sources (free fatty acids) to survive. Training in survival mode makes for less than optimal performances. Analysis of thousands of physiological profile tests reveals that after only two weeks of compromised glycogen stores, the anaerobic system can become unavailable. When a System shuts down, the athlete's ability to produce the same bioenergetic outputs will be physiologically unattainable. The shutdown of a System will directly correlate to an athlete's ability to perform the same event at the same speeds or performance levels.

Therefore, the concept of eating low/no carbohydrates to create "metabolic efficiency" to enhance performance defies both the First and Second Laws of Thermodynamics as well as challenging Carnot's Efficiency Theorem along with the Grady Human Performance theory.

To review, the First Law of Thermodynamics states that energy output is unable to exceed the input. Therefore, the concept of "metabolic efficiency" challenges this universal law of thermodynamics and human bioenergetic systems by merely requiring the output to be higher than the input. It is impossible to become more efficient when the output (ATP expenditure) exceeds the input (ATP synthesis). Efficiency refers to how well an energy conversion or transfer process takes place with as little loss of energy as possible. Energy efficiency is qualitatively measured by the ratio between OUTPUT and INPUT energy. In the case of "metabolic efficiency" principles, the efficiency of the human system will end up being higher than 100%, which is thermodynamically and bio energetically impossible.

As referenced previously, Carnot's Efficiency Theorem is based on the Second Law of Thermodynamics which only applies to machines where fuel is burned. The law also only applies to human "engines" where fuel is burned, again where the "metabolic efficiency" concept goes awry... NO fuel, NO efficiency.

Some endurance coaches have proposed the thought that low glycogen stores, caloric restriction, and fasting while training is the signal for an "adaptation" toward increasing fat and reducing carbohydrate combustion at faster speeds. The problem here is that they are taking a small piece of information and making shortsighted assumptions to suggest an adaptation toward increasing fat-burning and decreasing the reliance on carbohydrate-burning can be enhanced by reducing carbohydrate intake and increasing fat intake. Instead, denying an athlete's body of essential glycogen molecules needed for glycolysis, electron transport, and oxidative phosphorylation will undoubtedly limit rather than enhance performance gains for endurance events. These lackadaisical assumptions, along with negative performance correlations, will also cause athletes to suffer potential Bioenergetic Deficit and a myriad of other long term detrimental physiological effects that will have harmful health consequences. The only adaptation that occurs when foregoing the use of glycogen on long endurance or high-intensity sessions, which are known as glycogen depleting sessions, is that one can become more reliant on free fatty acids as a fuel. But, as stated earlier, free fatty acids reform ATP much more slowly than carbohydrates so this adaptation may only for slower races such as ultra-marathons. Thus, training while glycogen depleted reduce an athletes ability to train at higher race speed and, in doing so, will undoubtedly create a Bioenergetic Deficit and limit metabolic adaptability.

Chronic caloric and carbohydrate restriction adaptation is a survival mechanism to a bio-energetically inferior source of fuel, fat, that is less than optimal for improving performance will likely have a negative impact on metabolism.

Those striving to achieve "metabolic efficiency" inherently mean that one's metabolism is conserving energy, but unfortunately for those trying to lose fat that energy conservation comes in the form of fat storage. Although most who practice these methods experience some initial weight loss, they are most certainly failing to increase overall metabolism. It may seem almost counter-intuitive, but restricting carbohydrates, and relying on fats as the primary fuel source will increase fat storage.

While intentional under-fueling may have some short-term exercise or health benefits for the sedentary population or for recreational fitness advocates, under-fueling in athletes will lead to underperformance 100% of the time. Chronic under-fueling will also lead to a myriad of health consequences associated with the inability to maintain homeostasis, including but not limited to reduced hormone production, underperformance, injury, depression, and more. There are no shortcuts in the laws of ergy for natural athletic performance improvements. Energy production and usage come at a cost, and restricting carbohydrate consumption will adapt the body to an inferior bioenergetic source.

As a beginner or shorter-event athlete, it may seem unnecessary to fuel during training because workouts are short or low in intensity. Still, adequate fueling every single day will enhance training responses and energy levels for long term success. Different physiological testing protocols can be administered on an athlete to determine the amounts of net blood lactate present when performing an activity at various lengths of time and intensity. The net lactate output has a direct correlation between substrate and energy system utilization. As the intensity and time of exercise increase the reliance on blood glucose and muscle glycogen increases. These substrates are necessary to continue activity at the same work rates and net lactate outputs. To achieve the intended physiological response and maintain or improve system processing rates, frequent fueling during training is absolutely recommended.

Is it possible to complete long or intensive training sessions without fueling? YES. Is it wise to complete long or intense training sessions without fueling? NO! For example, a "long-run" of 10 miles is scheduled for Sunday morning. Most are unable to complete this "long-run" under 60 minutes, but many can complete it somewhere in the 60-75 minute time range. If you simply wake up, get dressed, and head out for a run without ingesting some food and water, but you will not achieve the optimal physiological responses that will improve your performances in competition.

While running along, one may notice they feel good, the pace is increasing, and the run is completed without a problem. However, there is a problem, and many athletes see this in their training regularly. When running in Aerobic Foundation and the PAC zones heart rate will shift higher into another energy system to maintain the same pace when fuel runs low. If properly fueled before this energy shift, energy system utilization will be maintained, and paces will continue to increase within the same heart rate zones. Additionally, the intended physiological response will be achieved by staying in the proper system for the training session.